The author

It Is This Way With Men Who Fly

By Frank K. Thomas

BJW PRINTERS
BECKLEY, WV 25801

1993

Standard Book Number 87012-311-4
Library of Congress Card Number 78-52774
Printed in the United States of America
Copyright © 1978 by Frank K. Thomas
Fayetteville, West Virginia
All Rights Reserved

1978-1984-1989-1991-1993
Fifth Printing

Contents

Preface xiii
Acknowledgments xxi
Jim Learns to Fly
   Jim Learns to Fly 3
   1945 6
   Jim 6
   No Apology 9
   Junior 9
   Aircraft Down 10
   The Flight Instructor's Pledge 10
   J. H. Page 10
   A Letter From a Student 11
Flying Stories of West Virginia
   Miracle Or Coincidence 15
   How the Judge Got His Tail
      Feathers Wet 18
   In the Beginning 22
   Long Trip to? 24
   Panic Button 24
   Harold Stewart 25
   Survey My Airport?! No 25
   Sam Mead and the Cat 26
   X-Rated 27
   Ambitious Student 28
   PA-20 Piper 29
   Help Wanted 29
   No Help Needed 30
   Harold Via 31
   Second Sight 32
   Gas Boys Don't Give Clearance 33

Why the Landing?	33
I-89	34
Whip Stall	35
Airline Or Covered Wagon	35
Thieves	35
Attempt to Kidnap	36
Surprise	38
Taken For a Ride	40
Fayette Airport	41
Air Force Bases	41
Dixie	42
Fire in the Woods	43
Airport Ghost	44
Clark Field	46
Have a Seat	46
Smooth Landing	47
Snake	47
Snow and More Snow	48
Snow Go	49
Six Degrees Below and Getting Colder	50
Hurricane Camille	51
Insurance	52
Morris Burnstin	54
The Uninvited Guest and Squadron Three	55
The Cessna 120	56
The Cost of Buzzing	57
Prop Wash	58
Propellers	58
Airplane Prop Kills Businessman	59
The Big Chase	59
Why the Sight-seeing Trip in 1942?	59
Fun and Foolishness	60
Raborn Cook	61
Thanksgiving Day	61
Gambling in 1945	62
Sound Plane	63

Freezing on Controls	63
Deceitful	64
To Fly What?	65
Emergency Message	65
Fayette Airport	66
Handbills	67
GI Bill	69
Bad Checks	69
Jennings Wagon	70
Paul King	71
Fell on Me	72

Flying History
Jack Brown	77
Up Yonder	78
Captain V. I. Kessler	78
The Father and Son Team	79
Roy Swanigan	79
Wheeler L. Weikle	80
For the Record	80
John Frazier	80
Hobart Booth	81
National Fight for Flight, Inc.	82
West Virginia Aviation Association	83
June 28, 1960	84
Morris Raymond "Dinger" Daugherty	85
Daugherty	85
Charles E. Yeager	88
History	88
Archie Clemons	89
For the Record	90
James K. Turbo Turley	90
James Turley	91
L. S. Scott	91
Mercer County	91
Glen Clark	93
I Remember James H. Tolley	94
Jim Tolley	95
Charles Lilly	95

Gill Rob Wilson ... 96
Robert Thompson ... 97
Staff Sergeant James K. Hall ... 98
Howard Clifton "Tick" Lilly ... 99
Lady Flyer ... 99
Collapse of Hangar's Roof Under
   Snow Crushes Plane ... 100
West Virginia's Most Experienced
   Pilot ... 100
Greenbrier Pilot Wings Back
   Into History ... 102
Godsey ... 104
That's My Car ... 105
I Remember H. L. Sessler ... 105
Foggy—1941 ... 107
White Sand ... 107
Bamboo Bomber ... 108
Retrieve ... 108
February 9, 1969 ... 109
Lewisburg—1949 ... 110

For the Record
   For the Record ... 113
   Weather ... 113
   To See Or Not to See ... 114
   C. E. Martin ... 114
   July 2, 1942 ... 114
   April 8, 1951 ... 114
   August 10, 1968 ... 115
   November 14, 1970 ... 115
   Without Wing Tips DC-3 ... 115
   Bail Out ... 116
   Dope ... 116
   Night Riders ... 117
   Twelve-year-old Saves Two
      in Plane Crash ... 117
   This May Be a First ... 118
   Tell Us Why ... 118
   November 20, 1948 ... 119

February 10, 1951	119
February 11, 1951	119
Registered Mail	120
Gifts	120
Rainelle	120
From the Scrapbook	121
Water	121
From the Records	122
From the Record	122
Car Gas in Airplanes	126
They Will Fly Without You	127
South of the Border	127
Low-Reckless Flying Costs Pilot	128
Hail	128
1956	129
For the Record	129
1958	130
Ambush	130
April 27, 1963	131
Fayette Airport	132
40 West	132
This You Should Know	132
From the Files—Charles Yeager	133
From the Files	133
1951	133
Why Not Under the Bridge	134
Legal at Last	134
'Lost' Navy Planes Circle Over Beckley	134
October 27, 1973	135
Timing	135
Flight Breakfast	136
Flying Priest Makes Error	136
Fayette Airport	137
Newspaper	137
Lost Planes	137
Heaven Or Hell	138
The Reverend W. Jones	138

For What?	139
Power Lines and Low Flying	139
Low Flying	141
Wives	141
Dust	142
Men	142
Work	142
Morgantown	142
Never Make It	143
True Humor	143
Flying Bug	143
The Worst of All	143
Wind—1946	144
Wind of 1968	145
Wind of 1947	146
Tornado	146
Strongest Winds	147
Carson Green	147
Gas Thieves	147
Wind	148
Desperate for Office	148
Wind—1949	149
Wind—1970	149
Thunderstorm	150
Gas Crisis	152
Gas Thieves	152
Log Book	153
1968	154
Lost Plane	154
Rich Taglang	154
Instruction	155
Who Is Lost?	155
Thought	155
Christmas	156
September 25, 1973	156
March 1974	157
Silver Wings Fraternity	157
Twins	157

x

Pylons 157
Engine Failure 158
Sportsman No 158
Plane Crash Kills Three 159
Low Flight 159
Unidentified Flying Objects (UFO) 159
Unidentified Flying Object Sighted
    Near Beckley Airport 163
UFO 164
Phenomenon 165
UGO 166
Permits 166
1961 166
1959 167
April 4, 1974 167
November 4, 1975 167
Thanksgiving Day—1977 168
He Sent Me Back 168
July 1967 171
Final Report on Accident 172
Meanderin' Around 173
The Last of the Little Strips 177
Forced Landings 182

The Battle With the Bureaucrats
    1952 185
    More Thunder 185
    How the Inspectors React 186
    From the Records 187
    Monthly Inspection 188
    PA-18 Radio 189
    New Aviation Group 189
    Conflict 191
    Controversy 191
    State Aeronautics Group Blasted
        by F. K. Thomas 192
    Why No County Airport for
        Montgomery County? 193
    End of Oak Wilt 195

Tax 195
Tax 199
Letter by G. L. Massey 200
Thoughts of 1978 202
Alpha Or Omega 203

Preface

This book—its short stories of the experiences of others and myself—with those that have known flying success, failure, sadness, joy, tension, anger, body pain, mistakes, carelessness, elements of weather, the socialistic conspiracy against small business, and worst of all, the harassment of the armchair bureaucrats.

With all of this, there has never been a moment of doubt that flying was meant for me. May I justify this statement by saying that I am one person that is where he wishes to be, doing what he wants to do. I would not trade with any man. I have been through it all. At times I am tired, but not of flying.

I have tried to walk and fly honorable in the vocation in which I have been called. With a will to continue, a never failing faith in God, I respectfully submit *It Is This Way With Men Who Fly*.

Frank K. Thomas

IT IS THIS WAY WITH MEN WHO FLY

This poem was written and dedicated to Wheeler Weikle for his accomplishments in flying.

When dark and restless skies are passing by
Like the clouds, I too am restless to fly,
To see the mountain's other side.
Onward, upward, it beckons me until at last I must
 surrender homeward bound
To the bonds of Earth, compelling man's spirit to look up
Henceforth, every landing a vow at dawn or dusk I will fly
 again
It is this way with men who fly.

Frank K. Thomas

This poem was written in memory of Charles Ramsey who crashed in 1962 near Logan, West Virginia. He crashed in a light plane while doing forest work for the state of West Virginia. His final words were, "Father, I must fly."

FINAL FLIGHT PLAN

Out of the valley of the shadows, into the clear blue sky
  Father, further, faster I must fly.
A flight plan for the future I have filed, not on victor
    but on victory,
  Further, faster I must fly.
Out past the moon, stars and Milky Way, thru the Gates
    of Pearl, down the Streets of Gold
  Further, faster I must fly
To a landing in the Master's green pastures
Then before the Great White Throne I knelt to close my flying
  career.
He said to me, "My son, did you think we have no need for
  flyers up here?"

Below are a few lines I have written expressing my feelings of our state's loss.

LIFE'S SADDEST HOUR
November 14, 1970

They were the best Wichita and Marshall possessed.
You say you'll never understand.
Stop! for a moment and say a prayer.
Know that there is one who cares.
Look about you. There is one that will not doubt you.
Listen with your heart—even though it be through tears
How he had loved them all through the years.
We must know and know it well
That the great Coach has called them to Heaven to dwell
It is our loss—the Master's gain
I believe that Marshall and Wichita are scheduled for a game.

Acknowledgments

Some of these stories have been previously published but where permission to reprint was needed it was granted: Miracle Or Coincidence, by Frank Thomas, The Conquest-Regular Baptist Press; Why No County Airport, *Montgomery Tech Collegian*; The Last of the Little Strips, by William Blizzard, *Sunday Gazette-Mail*; Meanderin' Around, the *Fayette Tribune*; Oscar Tate, by Adrin Gwin; Greenbrier Pilot Wings Back Into History—White Sulphur Springs—Red Baron, By Dave McCorkle of the *Charleston Daily Mail*; Daughtery— From the *Wheeling News*; Letter to the Senator, by G. L. Massey, permission was given by Mr. Massey to use this; How the Judge Got His Tail Feathers Wet, by Judge Charles L. Garvin, Jr. This was written by Judge Garvin to use in the book.

NEW RIVER GORGE BRIDGE LONGEST OF ITS TYPE IN THE WORLD

SIGHT-SEEING RIDES ANYTIME
FAYETTE AIRPORT

FRANK K. THOMAS, Manager
Phone 574-1035

Jim Learns
to Fly

JIM LEARNS TO FLY

His name was Jim. He was a real person, growing fast in body and mind. He was timid and trying hard not to show it. Most of the time, he rode in on his bicycle, but sometimes he walked or hitchhiked.

He said, "I think I would like to learn to fly. I have a newspaper route at the *Beckley Post-Herald,* or it may be a hundred other jobs. I can only fly of the evening." There was a gleam in his eye as though he had stepped into a new world. He said, "I know Bill. He told me all about it. When can I start?"

As I looked at this bundle of enthusiasm, eagerness, alertness, impatience for success for the day he would be up there on his own, it was easy to see that there would be no discipline problems. Only one—trying too hard.

As we walked together to preflight the airplane, my thoughts went back to a hundred other Jims, most of them the same. But there is the other type kid. Big Daddy, rich and tough, leads a young, scared, self-conceited young mammy's boy into the office. Big Daddy, a self-made man, working hard in timber, coal, or gas, is sometimes worth a million. He, Big Daddy, sleeps in the front of his truck to save time out on the job. He is the rugged individual of the hills, who, until now, had no time for his son. He was leaving the raising to the boy's mother.

"Mr. Thomas, this is my only son, Persey. Take him out there and teach him to fly. It will help make a man out of him. Give him a couple of hours; we came a long way," he replied. "Two hours is too much," I commented. Daddy says, "I got the money if you got the time." It would have been less painful, less frightening to Persey if it had been a major operation.

"Let's go fly, Persey," I said. I knew there must be patience and understanding. Persey said, "Could we just ride this time? I have never been in a small airplane before. This is Dad's idea. I frankly don't want to fly." Off we went. I thought, If I could only relax him, win him, show him how to make turns. If he understands where the nose and wings are in straight and level flight, I will have earned my money. Persey was too scared and nervous to grasp the controls. "Relax Persey and we will sightsee for a little while," I said. After twenty minutes he said he had butterflies in his stomach and asked if we could go down, but he didn't want me to tell his Dad that he had gotten sick or scared.

Upon landing, I told his dad that I was not feeling well, and we would try another day.

Jim was back to try again. He had a book from the Fayette County Library and as we started our run-up before takeoff, he had so many questions that I asked him if he could save some until we returned. He was so eager to grasp all he was shown. I myself hated to come down. Being up with Jim was the best part of the day. Landing with Jim seemed to be the shortest thirty minutes one ever experienced. Jim asked how long before he could solo. I wanted to tell him that he was above average, but he, I was afraid, might become overconfident. Or he may, as many students do, reach a plateau of learning, or even a rut, which may take two or three hours of flying before his rate of learning would start progressing again.

Then there was the Jim student bearing long hair and dirty, with that sweat and charcoal smell of dope. "Buddy," he said, "I'd like to take you." I didn't tell him where I would like to take him. He was told that I had so many students that I could not do justice to the ones I had.

We were told in the Federal Aviation Agency (FAA) Aircraft Owners and Pilots Association (AOPA) Seminar that we were not to take that long-haired, bearded crowd. This was told to us by the FAA psychiatrist doctor, who said that they have mental problems. I followed this for some time until it became so popular that the flight instructor simply had to surrender to the inevitable (especially when a long haired youth said, "Look whose picture is on the money we pay you. How long is his hair?").

Then there is the girl that invades our male domain, wishing to fly. This is usually a rough job, but not always. It may be that she will be as dedicated a pilot as any man. Most feel offended when you, through necessity and safety, tell them they must keep the nose down and airspeed up while gliding or it will stall and kill us both. Some will cry when told.

Stepping out of the office I see the controls moving on the airplane. The time draws a little closer for Jim and his third lesson. "I thought I'd look it over first," Jim said, "and will you explain density altitude and its effect on the controls?" "Yes," I replied, "if you are old enough to ask you are old enough to know."

We were off for number three—allowing Jim to taxi the airplane on the ground. He swerved from one side to the other, before we reached the takeoff end of the runway three, which meant the airplane would be heading thirty degrees on the compass upon takeoff. I told Jim that after check out we would try it together, the instructor being able to override the controls on small mistakes. I told him, "Jim, you must be careful and not touch those brakes on the take-off. This will pull the airplane sharply, dangerously to one side." He did anyway. With a quick, clear tone I said, "Get off those brakes." He did.

Once off the ground the controls are so much easier to handle that Jim forgot his mistakes and started practicing his turns and the coordination between rudder and aileron. "When time to land, Jim," I said, "you may try this landing." You could see him tighten up on the controls. He was very responsive to verbal instruction. "Now as we become closer to the ground, pull back slowly on your controls," I told him. As the airplane touched the ground it rose again some six inches. Jim, being tense, thought it went six feet.

While practicing landings with Jim, a few hundred feet from touchdown, there was something on the runway. Jim applied full power to the airplane and this was a good practice round. The obstruction on the runway was a small horse or pony.

I picked up the microphone and called the office to ask the

boys to get a board and hit that critter on the rear and run him so far down the road that he would never come back.

The same day one of our local flyers flew to Beckley. He heard the message "run him off," and they wanted to know who Frank was running off.

1945

Before Jim there was Jean. Jean was Mrs. Jean Wise Crouse, the first person to solo an airplane within the boundaries of Fayette County. Her charm, beauty, and enthusiasm added much to the little flying school at Fayette Airport.

JIM

Jim's mother called and asked if I would hear her story. She first stated that her son must not fly. Her statements that followed revealed that her life since December 6, 1946, had been a nightmare. The story involved the most mysterious flight in history. It involved the disappearance of her only brother, thirteen other men and five airplanes. Then only a few hours later another aircraft with thirteen men disappeared without a trace. Twenty-seven men vanished.

Around 2:00 p.m. at Fort Lauderdale, Florida, United States Naval Air Station, five planes with a total crew of fourteen departed for a short routine air patrol flight. Over one-half of the pilots were highly experienced. This flight was to be of two hours duration. The weather was perfect, wind a sea calm, and visibility unlimited. Then, one hour and forty-five minutes later the flight leader called the tower declaring an emergency. The tower asked, "What is your position?" The answer, "We are confused." They were told to head west, but they didn't even know which way west was. The flight leader turned his command over to another pilot. Immediately a Martin flying boat with thirteen crew members and all the emergency equipment known to men, departed to search, never to be heard of again.

The next day, hundreds of planes and all the available ships speeded to the area, broadening their search for weeks. These seven planes with their twenty-seven crew members simply flew into the unknown.

Jim's mother, "Now you know why Jim must not fly." I replied, "I am in full sympathy with you. Your loss was great."

There is another story of Ted. His mother forbade him to fly. Weeks later he was killed in a car accident.

Weeks passed and Jim did not show up. His mother called often to ask if we had seen Jim. We had not. There were nights he did not come home. Then Jim's mother called, "I have changed my mind. Jim can fly if you will take him. He has been running with a bad crowd." I told her that I would be very happy to take him, to send him out and we would pick up where we left off.

With such a wish for flying, I felt it would be hard to counsel Jim on his morals, but it must be done. When his eyes caught mine, it was one of those rare experiences which we all have had . . . a full message of respect for each other. There was no need for further counseling. He had returned all the way back to normal. Jim now had the full respect for his parents that they had for him.

Hundreds of landings, stalls, spins, go arounds had been shared between us. I told him, "Jim, if you do this good tomorrow I will solo you."

The next day as I was sitting in the office, I heard several cars come in—eight in all. Jim's mother had led the parade to watch Jim solo. All of Jim's aunts, uncles, and schoolteachers had been invited to see the great event. This was bad. There should be only Jim and myself or those who usually came with him. We would see Jim preflight check the plane. After the first landing I knew it would be a disappointment to terminate the lesson until better times. It was my duty to tell Jim's mother. She said, "You don't know Jim. He does not care if all these people watch." "Nevertheless," I replied, "send him alone tomorrow. I will send you a picture of his first solo." There was some talk from the crowd, wondering if they would take him to another instructor at another airport, and if they could see him solo today.

The next day Jim showed up for the first time a little shy. I said, "Today this will be only routine. Jim, take that airplane and do me two nice landings just like all the others you have made with me." Jim did. I said, "Congratulations! Now you are a pilot. This word belongs only to those who have made their first solo flight. You are even better than most who learn to fly from a very small field. This is only the first step of your learning. Most likely, the greater thrills are yet to come."

Each lesson added to Jim's skill. For his own protection before we allowed him to fly alone we gave him a short check ride. Each flight allowed him to fly a little further. We call that weaning them.

Jim told me that he had saved a little extra cash and wanted to start on his cross-country. Accompanied by an instructor on the first cross-country, Jim prepared his course with all the check points and time and distance. It was to be to Blacksburg Virginia Polytechnic Institute (VPI) Airport. Jim, with a little overconfidence, let me know he did not need my help. I then asked him at what point, if he should happen to get lost, did he want help? He replied, "Let me find the way." I asked, jokingly, "What if we see the Washington Monument?" He replied, "No chance."

In thirty-five minutes we were six miles east of Blacksburg. In ten more minutes the city of Roanoke was beneath us. Jim reported to me, "There is Blacksburg." We landed and pulled in front of the terminal. Jim said, "Look there. This is Roanoke." Jim then asked me if I knew it was Roanoke."

Weeks passed with Jim flying regularly and asking after each flight, "When are you going to allow me to fly to Charleston alone?" The weather was bad with visibility poor. We call poor visibility in Kanawha Valley "Rockefellow Perfume."

I believe Jim thought he would never be permitted that flight alone. But the day did come. Jim came to fly and I asked him, "Why do you never fly to Charleston alone?" Jim, finishing the flight, thought that that day he had graduated from the air force academy.

NO APOLOGY

I make no apology for giving as many check rides as I do to these pilots that have not flown for a long period of time. Suppose I give one-thousand check rides that are not needed and one that is—it would be worth it all.

Arriving at the little strip was an air force man with those shining eagles and a beautiful set of wings. One would suppose that he was the best of flyers, and he probably was. He asked to rent one of the little birds and I explained, "We will ride around the field once and see how you do." He was insulted.

He was then told that without a check ride he stood as much chance of getting one of my Cubs as I would flying *Air Force I*—the president's plane.

JUNIOR

The year 1942—it was my first year of flight instruction. This was at Beckley-Mount Hope Airport. For Christmas, H. L. Sessler gave me a flying outfit. It was the flying style of the day. It was similar to the flight officer's uniform—olive green shirt, peach-colored pants, with a tie to match. I added to this outfit a leather jacket, white scarf, General MacArthur sunglasses and short boots. These items were hard to come by in those days. I was so proud of them I hated to take them off at night.

My first student each day was a little black boy about sixteen years of age. The boys at the airport gave me a hard time, calling my student Frank, Jr. I did not care—that is, until Frank, Jr., walked in with an identical carbon copy of my outfit.

I could hardly wait to get this kid up. The lesson would be spins. To my unbelief, after the fifteenth spin, he was still with me. I was sick, but not Junior. I had to surrender homeward bound to try another day.

Somehow Junior did not get the message. He flew with me until he was flying solo. His parents moved to Chicago and Junior went with them. Free at last—free at last.

AIRCRAFT DOWN

James Price, flying a Piper Cub in the Fayetteville area, encountered a strange noise about the airplane. Believing this may be close to doomsday for him, he headed to the nearest green pasture, which was a cornfield. He landed safely and called the airport.

We found him and the airplane safe. The trouble was found to be a broken tachometer cable, whining in its housing. The seriousness of this matches a broken speedometer cable on a car.

We soon learned we had destroyed some corn. The damage was eighty dollars. Our insurance company rejected the claim, asking how many bottles of corn we had broken.

We were sued. We lost and our insurance company paid the justice of the peace court justice for the plaintiff.

THE FLIGHT INSTRUCTOR'S PLEDGE

Wherever I fly I pledge that my first thought will be for the safety of my student and those persons about us; to further the student's knowledge for the betterment of their future and our country; to do my work with patient understanding and pride.

J. H. PAGE

Jim Page, seeking a private pilot's license, applied for an appointment for a flight check with the examiner at Charleston.

He was told it was necessary for one of their instructors to recommend him. They had not even flown with him yet.

Jim, anxious to obtain his license, submitted to their request. It was winter and the days were short. Jim made several trips, returning after dark. This accomplishment most commercial pilots would not attempt on such a small strip.

A LETTER FROM A STUDENT

Dear Frank,

It was truely a pleasure to see you again, to see you in such good health and flying to your hearts desire.

A week before I visited Beckley to see my parents, I sat here in my study and thought about the people who influenced my life the most and who gave me a chance to be somebody in life. You will never know how much you have influenced my life in the short time I was around you.

Frank, besides my father, you have had the greatest influence, the reason being flying is my first love and you made it possible for me. My first flight with you was on April 4, 1971. To get back to the point I had very little money, and my father could not have paid for my lessons without straining the family budget. But, with your kindness you gave me a summer job, with the work I did it still could not have paid for my total bill for flying. In reality you gave me my license! I will always be in debt to you, if ever you need help it is just a call away. (I mean this Frank)

While I was around the airport I also picked up a business sense that will be valueable to me later in life. There were two main things I picked up from you (1) That happiness is not getting all the money just for the satisfaction of having a surplus but, helping others when they are in need. (2) The other you told me when we were mixing concrete and it was something to the affect: "that learn everything possible when your young because you will never know when you will need it."

My first love is flying and someday I am going to have my very own airport so I can help those that could not fly because of a lack of money, you gave me my chance now I will give them theres.

BRUCE Mc DANIEL

Flying Stories

of West Virginia

MIRACLE OR COINCIDENCE

I can do no less than share an experience far more rewarding than any mortal man deserves. This event touches simultaneously the lives of too many to be a coincidence.

One morning in early July, with a state aerial survey observer, I was assigned to fly the woodlands of West Virginia, in a small Piper Cruiser, in search of the Oak Wilt blight which threatened to disease and destroy our hardwood forests the same as the chestnut blight destroyed our chestnut groves shortly after the turn of the century.

The blighted oak trees are easily recognized from the air by the wilting, curling, and shedding of their leaves. This requires extremely low flying for positive identification. The pilot circles the area, and the observer maps the location of the blighted trees. Another low flying pass is made to throw rolls of adding machine paper or toilet tissue into the tops of the tree. This is to help direct the ground crew, which then destroys the trees to prevent the spread of the fungus disease.

It was one of these flights that gave William Grafton and me much to be thankful for when we returned safely. The airplane, in pilot language, was a "home-steader ground lover," which means that it takes a long run on the ground to become airborne and equally as long a landing roll to stop. But for its few faults, it is excellent for forest patrol because of its slow flight characteristics and low fuel consumption. This type of flying is often considered the most dangerous of all civilian flying. A low flying waiver permit must be issued by the Federal Aviation Agency for this flying.

We departed Fayette Airport, Fayetteville, West Virginia. Rising above the early morning fog which gave us a complete overcast, we flew twenty miles west without sight of the ground. Failing to find a cleared area we decided to fly southeast. Ten minutes later we were in the clear. Choosing a watershed to follow, we let down below the mountaintops into the narrow valley. This was soon never to be forgotten.

We were near the small coal-mining community of Tams, the most rugged section of the Appalachian Mountains. We descended even more to inspect some trees. At that moment the engine failed, 80 percent power out. Not even room to turn. There were three choices: highway, ball park, or railroad. Only seconds to make the decision. There was heavy traffic on the narrow highway. I decided the ball park was to be our landing field. It was small, but it would not endanger the lives of pedestrians.

Heading for the park, we were losing altitude very rapidly. Over our heads there was a high voltage power line which we did not expect; we cleared beneath the line. The ball park was so small that landing there could mean only one thing—a crash. Then the highway was clear of traffic except for one car coming toward us. The timing was good; we could clear the car and land on the highway. Then the unexpected, even more bad luck, the car saw us and stopped. It started backing up to try to stay ahead of us. Using the last of our power, we pulled over the car missing it by inches.

Most of the road was then behind us. We touched down within 175 feet of a curve in the road with a cliff of rocks on one side of it. Then the miracle of my life happened.

The plane was moving sharply to the left even though the controls were applied to keep it to the right side of the road. The hand of Providence was at work; the plane had found its way to the left, astride the railroad tracks, bumping down hard on the ties as it came to a stop. There was only slight damage to one side of the right gear.

Minutes later three ambulances came to the scene. We told the drivers we were sorry about that, but we had no business for them.

My co-worker and I looked the area over where the plane left the road at the only place where it was passable without extensive damage and probable injury to the occupants. We learned there was a truck coming fast around the turn. The truck passed within seconds after the plane cleared the road. After checking the controls of the airplane, we found them to be in order.

What caused the left turn that took the plane from the path of an unseen truck safely, gently astride the railroad rails? Miracle or coincidence, 10-4?

Back in our hometown a minister, the Reverend E. E. Hale gave testimony that he was working in his church at the time of our troubles. Great anxiety came over him. Deeply burdened for our safety, he knelt in prayer until assurance came that we were safe. He departed to his home. When he began to tell his wife of his recent anxiety, she stopped him. "Me too," she exclaimed. "Yes, brother Frank."

The conversation was shortened by the phone ringing. A friend was calling to ask the minister to listen to the radio. There was a news flash of the accident.

There are some things that will remain unanswered for awhile yet, but not to this country pilot. I must say that when I looked up from a sick airplane and saw we were clearing beneath a high voltage power line, a certain unmistakable, unexplained, warmness surged through me that told me all was well. This was not a coincidence, and this was not all.

My observer, William Grafton, was transferred the next day one hundred miles, hoping new country, a new airplane, and a different pilot would help him forget his experience.

Early the next morning he boarded a new aircraft to continue the work in the Allegheny Mountains. Near the highest peaks in West Virginia, the plane, in its first hour of survey flying, near the summit of a mountain, was caught in a downdraft; the wind pulled the airplane into a deep ravine where it came to rest with its tail in a tall oak, its nose on a cliff suspended between heaven and earth. A puff of wind came and pushed the plane well on the cliff allowing the pilot and observer to escape without injury. It took hours for the two men to walk out of the mountains to report their good luck. The plane remains there yet. The second accident the observer was in was within a twenty-four-hour period of the first.

Miracle or coincidence? 10-4. In forest language, "I understand. I have the message."

HOW THE JUDGE GOT HIS TAIL FEATHERS WET

By Judge Charles L. Garvin, Jr.

I was, at that time, a first lieutenant in the army air force as it was then called, and was stationed in Saint Joseph, Missouri, with the Ferrying Division of the Air Transport Command.

About March 14, 1943, I was ordered to report to what I recall to be the Ninty-Fifth Heavy Bombardment Group at Kearney, Nebraska. I vividly recall the first look at the "Flying Fortress" to which I was assigned as navigator. My pilot was a fine young major named "Cole," and a proficient pilot, he was, I might add.

My office was in the glass nose under the astrodome so that I had a marvelous view to either side, up or down, but not to the rear except upward to the rear. The craft was bulging with guns, turrets, etc., and gave the appearance of being invincible in war, though as you know, many were shot down.

Not long after I was sent to Kearney Air Force Base, the Ninty-Fifth was ordered to Rapid City, South Dakota. After a few days there we left a miserable snowstorm and flew nonstop to Gulfport, Mississippi. After a day or two there, we flew to West Palm Beach, Florida, and then to Borinquen Field in Puerto Rico. Thence to Georgetown, British Guiana (now called by another name). After Georgetown we flew to Belem, Brazil, and finally to Natal, Brazil, which is as I now recall about six degrees south of the equator, or about 360 nautical miles south of the equator.

I am not positive of the date, but it is my recollection that at about nine o'clock at night local time in Natal on April 4, 1943, our B-17, about twelve thousand pounds over gross, and with long-range tanks, ten souls on board, and our luggage in B-4 bags, set a northeasterly course across the South Atlantic

Ocean headed for Dakar, West Africa. There was a high, thin overcast which rendered the stars invisible, and I could not do any celestial navigation until about 0440 hours Greenwich time when I got a good shot at Vega. I remember this particularly because I was anxious about fuel until I could get a good shot which would enable me to determine our ground speed. I might say that I use the word ground speed advisedly when flying over the ocean. It seems a bit of a misnomer to me. I also recall that the shot of Vega encouraged all of us. Our ground speed was adequate. I was not able to get a good course line shot at another star, and so course wise, we were using dead reckoning, based on such information as we had. By daylight I began to use the sun to great advantage in my navigating. Dakar, according to my memory, is about fifteen degrees north of the equator, so the total distance from Natal on the East Coast of South America to Dakar, West Africa, is about 1,950 statute miles; a pretty fair hop.

At about 7:30 a.m. we lost all power in the number four engine. Oil pressure was needed to feather the propeller, and having none in that engine, we couldn't feather it, and so the windmilling propeller caused considerable drag and required much more power on the other engines and a higher rate of gasoline consumption on the other three. A dearth of oil in that engine finally caused a failure of some moving part in the engine which made the engine vibrate pretty bad, and in fact, set up quite a bit of vibration in the wing on that side. It was difficult to hold a heading and altitude.

Shortly after the trouble began we were ordered to throw our B-4 bags overboard. So we bombed the South Atlantic with ten or so B-4 bags. I was given the privilege of throwing out my own bag.

As we progressed towards Africa I asked the pilot to alter course a bit for Bathurst. It was closer than Dakar and was more directly towards land, which I might say, was much sought after at that time. I worked out a navigation problem on paper which allowed me to concentrate on taking successive shots with my octant and to check on our progress toward land. These shots, due to the vibration and difficulty in holding heading and altitude, I felt were not accurate, but an average could be struck, and it was better than nothing. You'd be surprised at the courage my being very busy seemed to give to the crew. They felt, they said many times later, that my navigation was bringing them closer to land, and I felt better about being busy too. I asked each man to take out a bit of navigation equipment with him, and the pilot asked each man to take out some food and water. The food and water were in one large container, and we felt that it was better to divide it up so that if injuries occurred in the ditching operation, some of the equipment and food and water would get out with those who were not injured, or whose injuries were not great.

Finally when we were about thirty-five miles from the coast of Africa, the pilot sounded the bell signaling his intention to ditch the aircraft in the water. We made our way from our various stations to the radio room with our parcels of equipment, food and water, and parachutes.

Shortly thereafter, though it seemed much longer, we made what had to be a very soft landing in the sea. It sounded like the plane was being torn apart. Radio equipment was torn loose from its moorings and went knocking about the room where we were piled on top of each other for safety. But I say it had to be soft in the sense of the word that this was a land base aircraft not intended for landings in the water and not built to stand that kind of stress. We were told that most B-17s broke in two upon ditching operations at or near the belly turret; this being a weak place in such landings. But our ship hit the water and skipped back into the air, and hit again much harder, but went back into the air. Then it seemed to strike with a great deal of force, and dug in, so to speak, and stopped. Our additional tanks, in the bomb bay were empty, and for that matter the others were almost dry. Major Cole noted that the entire prop assembly on the no. 4 engine came off and sank upon contact with the water.

Our plane floated for a long time. I'm not sure now how many minutes, but this was quite unusual. The pilot and eight of us promptly exited the plane onto the wing, and into two five man life rafts, which had been stored in the fuselage and which were ejected by the pulling of a lever, and then automatically inflated with, I believe CO_2 cylinders. We took our stuff with us including the parachutes. These came in so handy to protect us from the sun, and to catch rainwater. In addition, each raft opened one parachute so that it billowed out into the water to keep the wind from blowing us away from one man on board who got excited and dove out the waist gun window. After the ship came to rest he dove out and of course the wind had no effect on him, but it was blowing us away from him faster than he could swim and faster than we could paddle. We almost lost sight of him but the parachutes stopped us and enabled him to swim to us. This consumed about an hour. Our plane sank before we were able to get him aboard one of the rafts.

My feeling as we headed down for the ocean was that I would not survive. I don't remember actually thinking that exact thought, but in piecing together my thoughts later and trying to reconstruct them, I thought I must have felt that way, because I didn't mention myself in the prayer I offered to our Heavenly Father. I prayed for him to look after my wife and little girl back in Saint Joseph.

None of us were hurt. I cut my left a little, and a bottle of Chanel No. 5 perfume in my jacket pocket was broken which I had bought for my wife somewhere en route, and had failed to put in my B-4 bag. Needless to say, I was the sweetest smelling navigator in the air force for a while, and one of the happiest to be able to use my smeller.

I took a nap for a while and when I awakened, I got my octant, watch, etc., out and began to figure what was happening to us. We discovered that the wind and current was taking us back towards South America, and so out went the parachutes again. They make good sea anchors.

We also had a radio on board called a Gibson Girl, because of its hourglass figure. We had an antenna aloft carried there by a gas-inflated balloon, and we were not only cranking out the SOS on the Gibson Girl but our radio operator was keying out our position to any who might be listening. An American in Dakar was listening, and about three hours after we ditched he flew an old war weary DC-3 over us, and with a signal light told us that the British would probably get to us first to rescue us. Then he hightailed it back towards Dakar.

We were in the rafts about nine hours. Just before dark on that same day, which I believe was April 5, 1943, a British Short Sunderland used to patrol for Nazi submarines in the area flew over us and beyond, dropped his depth charges preparatory to landing in the water with his flying boat, and then landed towards us. Darkness fell rapidly, and he could not see us until we fired a Very pistol, I believe it is called. They taxied up to us, took us and our gear on board, and then did what I call a crash takeoff. He landed us on the mouth of the Gambia River, at or near Bathurst, West Africa.

The pilot of the Short Sunderland was from Australia. His navigator was from New Zealand. The British were hospitable and kind to us. We were well fed, and even clothed. I was given a pair of the best wool socks I ever owned, a pith helmet, a good shirt, and a pair of British tan walking shorts.

IN THE BEGINNING
January 1944

My enthusiasm was at an all-time high. I had leased Fayette Airport and had purchased several lots with timber, one tree being the tallest that I had ever seen.

Each morning I started to work with ax and saws. This was before I had ever heard of a chain saw. It was sheer joy to see each tree give way and crush to the ground from the blow of my ax, knowing there were several acres of timber that must be cut. Every lick with the ax brought the day closer when I would be landing my own airplane on my own airport. I had chosen this location, believing that it was so far out in the country that it would never be envied by any person and I would start a little flying school and serve the community as faithfully as I possibly could. As you will learn later, things do not always work the way you plan them.

After only a few days' work, snow covered the ground. I took my lunch box with me, because I was going to start early and was not to return until dark. This appeared to be a perfect day to burn brush. There were many hickory trees, and once started, they burn like coal. Shortly after the fires were started, I became sick. It felt as though someone had stabbed me in the side with a large knife. Feeling too bad to leave, I lay beside the fire, not cold, but feverish. The long day passed slowly. I realized those of my family would not know I was sick, but would think I was enthused and working late, and that it may be hours before anyone came to look for me. My car was one-half mile up a muddy road. I crawled at least one-half of that distance, reaching the car and being barely able to drive. To my delight, I saw a state policeman following close behind me and I had hoped that he would stop me. He did not. I proceeded home with much pain, reaching the driveway and calling to my mother, telling her I was sick. She asked if I needed help or a doctor. I replied, "Yes."

The doctor came immediately, felt my side, and said I had a ruptured appendix. The doctor walked straight to the telephone and called an ambulance. By that time, it did not matter whether I lived or died, the pain was so great.

The ambulance driver rolled me into the hospital and I was met by the chief technician. He asked where my slip of admission was. I explained I had none. He then ordered the driver to take me back where he had gotten me.

The technician had seen charcoal black on me from the burning brush and he thought it was coal dust. They were used to treating a coal miner that way if he did not pay into the hospital. The ambulance driver had much trouble getting the attendant to listen. When it finally dawned on him that I would be a cash customer, I was wheeled into the emergency immediately and the ruptured appendix was removed. I had a fourteen-day stay at the hospital, with good treatment.

This slowed the opening of the airport down by several months.

LONG TRIP TO?

Jack Wendell, Morris R. Reynolds, and John Witt rented Tri-pacer 1782P for a trip to Indianapolis, Indiana. It was Memorial Day and the big race, a must for sports loving fans.

This was before the days of good weather forecasting. The best radios that we had were VHT3, which we called Coffee Grinders.

This morning there was a heavy overcast, with one big break in the sky over Fayette Airport. Jack climbed through the broken overcast putting him over the top with clear skies and all bad weather below him. One would feel certain there would be a weather break between Fayetteville and Indianapolis. Reaching Indianapolis, there was no sight of ground. Neither Jack nor his passengers were yet instrument rated. This meant returning south and hoping for good weather. Three hundred miles back home; he was on his last hour of gas and time was up. The big hole through the overcast was waiting for him and looking through it was Fayette Airport. He had been to the races but all he had seen was Fayette Airport, blue sky above and clouds below—five hours and thirty minutes of flying.

PANIC BUTTON

I was flying over the Oak Hill area when I heard coming over the airplane radio, "May Day—096—Frank, where are you? This is Ferris and the engine has gone bad in 464."

All other aircrafts in the area ceased to communicate in order to give us an open, uninterrupted communication.

I asked, "Ferris, where are you?" He replied, "Over Gatewood, Frank. What shall I do?" My answer, "Check your selector valve on your gas tanks. Check your mixture control to see that it is full rich all the way in. Check your primer and make sure it is locked. Check your carburetor heat on and off to see if that makes a difference." His answer, "It's worse." I then told him to keep cool because sick engines would fly for hours then bring that bird in normal as always. He said, "I don't think I will make it. What shall I do!?" I wanted to tell him to pray, but he had always said that he would keep his place of business open one-half day on Judgment Day, so I decided I could forget that.

Then with surprise and pride he explained that he had found the trouble. It was the safety belt hanging out the door and banging on the side of the airplane.

He was not yet an experienced pilot and upon returning he wanted to go on a trip. "Frank, tell me where to go." I replied, "Not in front of these ladies."

HAROLD STEWART

Harold Stewart, a native of Kentucky, escaped across the border into West Virginia. At an early age, Harold was an excellent mechanic. He called the shop his kitchen. Harold informed every visitor who entered his aircraft shop that he was chief cook in his own kitchen and did not need new recipes.

Harold's shop was at Fayette Airport. He was finishing a fabric recover job on an old cub and the final touch would be to put the numbers on the wings. This was before the day of putting them on the sides of the airplanes. He had masked off the area not to be painted. Harold was a very neat person, but after his job was finished, the shop was a mess.

In walked the inspectors. Immediately they told Harold to clean up the shop. Harold took one look at them and I knew it was time to rush to the door and open it, for fear the door would be damaged. Out came the inspectors. Harold said, "When I said get, I mean go."

SURVEY MY AIRPORT?! NO

While landing with a load of passengers, as the airplane rolled to a stop it hit a large stake that had been driven in the center of the runway, doing only slight damage to the airplane. Checking other stakes and their markings, I found it was the same as those stakes belonging to Baker Engineering Company. Much mail exchanged hands between the state road commissioner and myself. Finally the commissioner said that it had not been sabotage.

Weeks later, looking down the runway, I found that the guys were back—stakes, transit and measuring poles. Not having yet recovered from my recent accident, I was using a cue stick for a cane.

I jumped in my little red Rambler. One of the men saw me coming and got away and the other started a running retreat. I had not realized that the ground was covered with a sheet of ice and a thin layer of snow. As I was trying to herd off the running engineer I approached him and tried to stop but the car slid right, then left. As the car repeated its action, the engineer ran right, then left, like a scared chipmunk. I sounded my car horn. He thought it was Gabriel's trumpet and slipped and fell, thinking his time had come. The car stopped just short of him. Out I came with my cue stick cane and asked him his name. He did not know. I asked him who he worked for and he could not remember. I decided that there was no reason why one did not know who those valuable engineering tools belonged to should keep them. He thought it was a good trade; to let him go in exchange for the engineering equipment. I, in return, told him if he remembered who they were working for I would gladly give them the tools. I wanted to sue them over the stake in the runway.

When they decided to survey around the airport for the road, Mrs. Mary Crisp, the owner of an adjoining farm, asked the engineers if they were going to take any of the Thomas property. They replied, "No. Mr. Thomas don't want us on his property."

SAM MEAD AND THE CAT

Sam Mead was ready to fly anywhere, anytime.
One day a call came to transport a cat from Lewisburg to Pittsburgh for a rich lady. The cat was placed in the back seat

of a Stinson SR-5. The cat was accustomed to riding in cars so it curled up for a nap.

Then the weather and scenery changed over Elkins from calm to severe with thunderstorms and lightning. The cat raised up, stretched, looked out at the great Monongahela National Forest, then jumped on Sam's neck and back digging his claws in tight to hold on for his ride through the thunderstorm.

Sam considered this an emergency—either he or the cat must go. Sam opened the small window on the pilot's side, pulled the cat loose from the back of his neck and attempted to throw the cat out. The cat did not want to go. He fastened all four feet on the frame of the small open window. When Sam would occasionally turn loose of the cat to straighten the airplane, the cat would then loosen one paw to scratch and bite Sam.

Sam said it would have taken five men to put that cat out, one to fly the plane, the other four, one for each paw.

Passing the thunderstorm area, the cat quietened down and wanted to cuddle up beside Sam. Sam grabbed it by the back of the neck and gently threw it into the back seat, hoping that it would break its neck. Then the cat began to lick Sam's blood off of its paws.

Upon landing, they were met by the lady and her chauffeur. The lady, seeing her cat covered with blood, asked Sam, "What did you do to my pussy cat?" Sam replied, "Not a x ! x ! x ! thing. That's my blood."

Sam, scratched, bleeding, and bitten, decided that he did not like cats anymore. In fact, since that time, I have never seen Sam without his pistol.

X-RATED

R. Harle, while flying the Oak Wilt Forest Program, trusted the line boy to check his oil. Harle learned at a late hour that the oil cap had been left loose, throwing oil out across the airplane onto the hot exhaust, creating much smoke. He was flying the Cabin Creek area near the West Virginia Turnpike, with his oil pressure low. He had to make a decision.

This area was full of mountains, power lines, strip roads and trees. There was a chance that the engine would hold out and there would be enough oil to make it to a small island ten miles away in the middle of the Kanawha River.

This island was located a few hundred feet below the plant at Alloy. There had been some flying from this island years before by Jack Fontis and Robert Blosser.

He landed safely, only to discover that there were no boats to take him to shore where he could call for help. So, he decided to swim for shore.

Finding a clean five-gallon bucket, he disrobed, put his clothes in the bucket, sealed the bucket, and started the long swim across the Kanawha. Halfway there he ran into real trouble. One of those large paddle wheel boats pushing coal barges came so close to Harle that he lost his bucket.

With no other alternative, he swam back to the island dressed as Adam without a fig leaf, and waited until rescue parties discovered him. Upon arrival, they found his face the same color as his red hair—also, other parts sunburned.

AMBITIOUS STUDENT

This story takes place on Harper Road, Beckley, West Virginia. It was told to me by F. M. McCaffey, Beckley, and C. M. Martin, Fayetteville.

In 1923, Mr. Martin possessed a plane which he used to fly to Beckley Harper Road Airport, then leaving an ambitious student to watch the plane.

Appearing on the scene was C. M. Goodwin, from Sophia, finding a young man watching the airplane. The youth responded to Goodwin's request for a plane ride, and told Goodwin all about the plane, including how to fly. They climbed into the cockpit and into the wild blue yonder they went. Things went well for a time, until the gas became low and darkness started creeping in. Then trouble began. Upon attempting to land, the first, second and third time they over-shot the field. The fourth try they ran out of gas. He hit a rail fence, tearing out about fifty feet of it, then went into the hillside. The passenger was unhurt, but the airplane was a basket case.

The youth collected his five bucks, admitted he did not know how to fly and was not supposed to take the airplane, closed his flying creel and took off by foot.

He has not been heard of by anyone since, believing he may have heard the call—go west young man. He probably became a cowboy.

PA-20 PIPER
1949

The first PA-20 that was shown me was at Saint Augustine, Florida.

I asked the owner about its performance and if he thought it would get off a short field, elevation two thousand feet. I told him the field was up hill both ways, with some mud when it rains and rough with obstructions both ways. He said, "It won't be long before there will be a helicopter on the market. Why don't you wait for it."

HELP WANTED

There is no time in my life that my spirit is so low and I feel so helpless to help mankind, as when an ambulance or police car pulls up beside my airplane with a youth on his way to a dope clinic. The question is—Why?

A few months before, the very brightness of freedom's holy light shone in their face. But now—dejection, loneliness, bitterness, and pain. Some of these youth are leaving home for the first time; with the feeling of being unwanted and shameful with crying and pleading for one more shot to make the trip; so strong, but yet so weak.

The finest looking youth of the community pleading and begging the doctor for one more shot to help him make the trip. This time, the doctor went along, stating he may become violent. He said, "If I give him one shot, he'll want a dozen. It is like a bottomless pit."

If it were possible that other youth could see this agony and torment, the hell of a bad trip, they would go forth as a living witness against the dope pushers.

Some weeks later, a call came to return the youth home. I was told that he did not respond to treatment. The caller said that the body would be ready anytime after twelve. Quite cold still, and pale and beautiful, I knew that the long dreary days and nights of agony had destroyed the youthful look but his pain was gone. Unclean, dirty needles had caused cancer. Dope sold by dirty men. My passenger was on his last trip home. We will handle him respectively, but what of those that are not yet hooked? Help wanted.

NO HELP NEEDED

A flight instructor assumes that when a person calls saying that he is a student with few hours and needs much check time, that this student will be easy to work with. Not so with our young engineer.

His story was that he only had four hours of flying time. As I started his instruction, I found his flying equal with any man's flying time with four hundred hours. I explained to Mr. Crutch that I did not know where to go from here. A man with four hours would not be capable of flying off such a small landing field. In searching for an explanation, I thought it might be possible that he had ridden in airplanes with others for hundreds of hours, which he acknowledged.

One week was to pass before his next lesson. I made inquiries. Was he a spy for another flying service or an investigator for one of the bureaucrats? I knew this was not the way the FAA operated.

I made a special visit to the FAA office and told the director of safety the story. He immediately told me Mr. Crutch was bad news. Mr. Crutch, among other things, had sold shares in an airplane, then fled town with the money. Later that month he returned in a Beachcraft under low, very bad instrument conditions. The FAA authorities were on the ramp to greet him. He told them he was the holder of an airline transport rating, and he did not even have a student pilot's certificate.

Before the next lesson I called him, explaining to him he did not need my help.

HAROLD VIA
Air Pioneer

Harold Via, a Beckley pilot, mechanic, and master of the parachute, is believed to be southern West Virginia's first legal licensed mechanic.

In the early days of flying an air show was not complete without Harold Via. He was sometimes accompanied by his daughter who did aero wing walks.

At an air show near Beckley, Harold's parachute failed to open. As the chute trailed behind him the crowd could see him fight the cords. This was a most tense moment. Via had fallen down and out of sight, as he struggled to loosen his cords. This was the last sight the crowd saw of Via.

The crowd rejoiced to learn that only a few feet from the ground the shoot blossomed. Via received a broken leg but he lived to fly and jump again.

This did not dim Harold's spirit. Mrs. Via said Harold's first love was flying.

Harold was also the builder of airplanes, flying those he built. His flying brought many interesting happenings. On one such flight he landed in a clover field at Adner, Virginia. A farmer chased him off with a shotgun, siccing his dogs on him. His airplane was safe until he could obtain legal help.

Airports were few in those days. There was another occasion of the same nature. Harold landed in a pasture field and was chased by a bull. He made it to the fence just in time to receive a little extra shove from the bull. The bull then returned and gored the airplane.

Harold Via and Dr. H. B. Wurtz purchased an airplane as partners, bringing it home from Indiana. Each was eager to try their skill at flying. Shortly after takeoff one pushed the button raising the landing gear. The other, unaware the gear had been raised, pressed the button again. Down went the gear and locked in place as though for a landing. As they returned they discussed why the airplane was flying so slow. Reaching Beckley, the button was again pushed to lower the gear, which retracted it. Thinking the gear was down locked for landing, on its belly it went, doing some damage.

Harold Via was one that was always ready to help. His love for flying did not include any thought for profit.

SECOND SIGHT
1948

As the work day began at the Fayette Airport, I felt an uneasiness—something I could not explain. Later that afternoon the anxiety grew. I asked one of the other instructors where one of the cubs was and who was flying it. When I was told, I stated, "He is in trouble. He is lost." The feeling grew stronger.

My closeness and love for this business sometimes gives me that which I called a second sight. For a time I felt much tension, then the uneasiness left. I knew the pilot and passenger were safe, though it was an hour before I heard from them. This is one of those things that cannot be explained.

The pilot, Harold Owen's, and his sister, Rebecca's, experience was as follows:

Harold had become confused as he was departing Pence Springs Airport. One hour later as gas was low, Harold landed in a very small clearing on top of a mountain in Clay County. How this was accomplished we do not know—a lot of luck and skill. There was not room to turn the airplane.

Nevertheless, he had told the natives I would be there in two hours and fly the airplane out. Climbing the mountain, I took one look at the airplane in the small clearing and complained, "It should just as well have been a crack up. We will never get it out." Harold's reply, "You did not even ask if I was hurt." I replied, "How could you be hurt. The airplane was not even scratched." The next day it took all hands to dismantle the airplane and carry it off the mountain piece by piece.

GAS BOYS DON'T GIVE CLEARANCE

A call came from the safety director of the FAA on June 26, 1974, asking if I was acquainted with the pilot of Cessna 7304G. My answer was, "Yes." Graham Pitsenberger explained that the pilot departed Kanawha Airport without being cleared by the tower. He asked if I would work with the pilot, Dr. Carl Adkins.

When I told the doctor of his mistake he asked if he could explain what happened.

He had taxied out holding off the runway, sitting for a few minutes trying to contact the control tower. Then a gas boy came running up and told him the tower was calling him to go ahead and take off. After departure he discovered he did not have the volume turned up on his radio.

I was practicing night takeoffs and landings with Dr. Adkins at Summersville. I asked him if he would like to try a takeoff and landing without the landing light on. He started his takeoff roll without lights, quickly snapping the light on. He stated, "Already I don't like it."

The doctor was asked, if I, Frank, would let you do what you wished with his airplanes. The doctor replied, "He will let you do exactly what he tells you to do."

WHY THE LANDING?

I had to make a quick flight to Ohio, near Gallipolis, to check on one of my airplanes that had made an apparent forced landing in a wet, soggy field. The ground was too soft for takeoff.

Arriving at the scene, I asked the pilot, "Why did you land here? Were you lost?" He shook his head no. I said, "Was the airplane out of gas or engine trouble?" "No," he replied, so I then said, "O.K. I give up. Why?" He answered, "Do I have to tell you. You know those large colas you sell. I drank two before I left and you have no rest room in your airplane." I told him that this had happened before, but other pilots had made better decisions. Mr. Coleman wet his pants.

One other case, one of Fayette's most prominent citizens filled a clean sheet full of water. At one time I carried an empty oil can, until I was cut on it. One of the students asked me if I cut my hand. No, it was not my hand.

One other student carried a milkshake cup. When filled, he tried to throw it out the window. The wind blew it back in his face.

All ye pilots take heed. All did better than my double cola pilot. A bath, a drycleaning bill and a Band-Aid are all much cheaper than dragging an airplane out of the mud, dismantling it and trucking it home.

I-89

Each airport has a number. Fayette Airport's is I-89.

Occasionally there is the big city dude lost, nearly out of gas, circling Fayette Airport several times, wondering whether to land or look elsewhere for a larger airport. He has been used to the runways five thousand feet or more. The big decision has been made. He will try to land. After all, he is a pilot, and the paper the FAA gave him says, "Pilot's License."

The miracle of his life has occurred. He did not kill himself or even damage the airplane. He steps triumphantly from his flying machine, making a few uncomplimentary remarks about our poor man's airport. He then asks where he is. His deep concern is if there is enough runway to take off on, asking if there has ever been an airplane like this landed here before. Our answer, "Yes, twice before. They went out good, both dismantled and put on trucks." He, being a brave soul, states he will fly it out. I then tell him, "Good boy. We need some excitement." It is my duty to tell him if he crashes in the woods the wildcats will get him before we could find him. This would be no worse than crashing in New Jersey and the hippies getting you. He departed and we did not receive a thank you note.

WHIP STALL
1946

In some cases the airplane slides backwards a short distance before the nose of the plane drops, whip stalls cause severe strains on the tail surface.

As a young instructor, through carelessness, I allowed the student to pull the nose too high. While practicing regular power stalls, the little J-3 whip stalled.

A fire extinguisher, not securely fastened to the floor, went through the skylight. The student, an older man, then lost his false teeth, which probably followed the fire extinguisher.

AIRLINE OR COVERED WAGON

A friend passed away near Miami, Florida. His remains were placed in the proper shipping case, then put on an airline and shipped via air for Charleston, West Virginia. Reaching Charleston, South Carolina, it was unloaded by mistake and the next day was put on another airline for home. Charleston, West Virginia, was fogged in. It was necessary to pass it by, and so it landed at Chicago.

The next day, Charley started on the long journey home. Someone shipped and again forgot to unload poor old Charley. He was now New York bound. Reaching New York, he was immediately reloaded into another airline and returned to Charleston.

Charley had broken all records of eluding the undertaker. Being three days late, the comment was, "We finally caught poor old Charley. He was a restless person who liked to run around." As I visit his grave, I wonder if they really caught him and put him in the grave, or is he out there running around somewhere.

THIEVES

For night protection of the airport, I occupied a room adjoining my office for sleeping quarters. The only other night resident on the airport was my dog, who only came to my room when there was a disturbance outdoors, either someone, a dog, or an animal. The dog pushed gently against my bed, never barking, when this happened and then I knew it was time to rise from my slumbers and be on my way.

On this particular occasion, on my way out, I picked up my rifle and cautiously moved to the front of the large hangar. I surprised three gas thieves. They were not youth, all were older than me. Two fled in the dark and the other stood as though frozen. With my gun pointed directly between his eyes, he considered it wise to stay where he was. I told him if he moved that I would throw his dead rear out over the Blue Ridge Mountains when daylight came. He replied, "But Mr., I ain't dead." I answered, "You mean not yet." His story was the old story. Here is what he said: "I ran out of gas and just pushed my car up in your driveway to get it out of the way. Those are not my cans and hose sticking in your airplane. They belong to those two who ran away. They were going to come by and pay you tomorrow."

He pretended to be very crippled and could hardly walk. As he came closer to me, it appeared so convincing that I relaxed and was off my guard. When he was in reach of me, he attempted to twist the rifle from my hands.

When he was admitted to the Oak Hill Hospital with cracked ribs, he told an attendant that a hit-and-run truck driver got him.

ATTEMPT TO KIDNAP

Darkness. Flying was over for the day, with only a few things to do before leaving. One of the explore scouts stayed to help lock up the airport and receive a ride home. We were doing some last minute checking in a back office when a car horn sounded.

I thought, "It must be someone wanting a trip tonight or early tomorrow." I told the young man, "You stay here. I will go see if I can help them."

Reaching the parking lot, there was a large, friendly man standing beside his car, with another man inside the car. He came over to greet me with outstretched hand and asked if I was Frank Thomas. As we shook hands, he fouled my arm into an armlock and told me to get into his car. The pain of my arm was almost unbearable.

With a loud shout for help, the young man answered. I asked him to bring the .22 rifle and he responded rapidly. I told the youth that this was serious and that the man was trying to kidnap me and that once away from the airport, would probably kill me. I asked the young man to shoot him if he did not turn me loose. The young man took careful aim and said, "Mister, you'd better turn him loose or I will shoot you." The man relaxed his hold—I stepped back and took the gun, then told the man in the car to stay put or I would shoot him.

I was not sure what I wanted to do with my captive, kill him or try to scare him to death. With the gun in his back, we went for a little walk. He cried, begged and prayed. To shoot him or not—that was the question. I then considered that such publicity would finish my ailing mother. I told the men, "Go and never return." He thanked me.

He had not been gone long before the phone rang. It was that man again. He swore he would kill me if it was his last act. I assured him that such an attempt would be his last and told him I would promote him to glory on sight if he ever set foot on my property again.

The next day I visited Roy's Loan and purchased a .38 Colt Revolver. The .38 was to be my constant companion.

The next day there was a threatening letter. Prior to his attempt on me, I can truthfully say I had never seen or heard of this man. My thoughts were searching past experiences trying to remember where or when I could have offended him.

A few nights later, as I was leaving Oak Hill, there was a car following me. When I would pass another car, so would he. When I speeded up, he was behind. Then when he grew very near he blinked his lights. He could not be shaken off. I thought if I could only reach Fayette Airport there the possession of my gun would be legal.

Reaching Maple Road, about one mile away, I would be on my own land. Then what I did to defend myself would be justified.

One-fourth mile to go, and the car was bearing down fast. Around me he went, forcing me off the road. Out of his car he came. I grasped the .38; there would be no more waiting. Did I have a surprise for him.

A certain calmness came about me with no more fear that he was lurking in some dark corner. Timing was important. The door of my truck cracked open. When he reached a certain spot, it would be time to step from the truck and move fast with rapid fire (for days I had been practicing moving and fast firing the revolver at five-gallon cans). I was ready. In a few seconds he would join his ancestors—that is, if he didn't get me first. For those that know guns, how I wished I had had a .45 Colt. After the first shot he would love everybody. I had confidence that he would wait until he was at the window of the truck which he would never reach.

My thoughts—Should I run, after shooting, to a telephone and call the police or back up, pull out, and let them hunt me, or would they ever find me?

The motor was running on his car. Time was running out. Two steps and he would be in eternity. Nearly point blank, it was then time to defend myself.

A shocking surprise when a loud voice spoke, "Frank, this is Trooper Lilly of the State Police. I am not in uniform and am driving my own car. What does a man have to do to get you to stop? You have no lights on the back of your truck."

My calmness was gone. The trooper said, "Why are you so nervous." I did not tell him. From that moment on I vowed not to carry a gun. I personally would rather be killed by some thug than to kill an innocent person.

SURPRISE
August 1970

It was an early afternoon in August of 1970. Frank Miller had landed a twin Aero Commendon on Fayette Airport. The flight was a combination training flight and visit to his father. The airplane was based at Fort Belvoir, Virginia. This airplane was once used by President Dwight D. Eisenhower for his short commuter trips.

As Miller departed, rising a few feet in the air, one of the engines failed. The airplane appeared to sink into the heavily wooded area beyond the airport boundary.

It was time for action. I rushed to the telephone telling the operator of the crash and asked her to send a fire truck and an ambulance. Then jumping in my car to reach the crash point, I drove to the south end of the airport. Due to heavy brush and trees I was unable to find the wreckage. I returned to the airport. As I was preparing a search airplane, before my eyes the most beautiful, pleasing sight I had ever seen, Miller and the co-pilot were walking out of the woods and over the hill, and both were unhurt. The airplane was a total loss.

Off in the distance was the sound of many sirens—fire trucks, ambulance, police, and what seemed to be the entire population of the town.

The townspeople swarmed around the airplane. I asked the state police to remove the public from the airport. The officers answered that I must make a written report. I told the officer I would put in an emergency call to the governor for help. There was one-hundred-gallon high octane gas spilling on the ground and if it went by a spark or a cigarette, dozens of people would be killed. As I was starting to the telephone to call the governor and ask for help, Deputy Sheriff Leon Ford, who is also a pilot, told the state police, "Thomas means it. He will call the Governor." The state police evacuated the airport except for the Oak Hill Fire Department. They were keeping watch until the National Guard relieved them. The Oak Hill Fire Department sent Fayette Airport a bill for $250. We turned the bill over to the army constable, Woodrow Bennett, who guarded the front gate of the airport with such a good job that if I would have gone out, I am not sure I could have gotten back in.

TAKEN FOR A RIDE

Fayette County had its own candidate for Congress, the candidate being Frank Love. Mr. Love asked me to fly over his congressional district with the airplane which had the powerful public address system. The district reached from Fayette to Harrison County and as I returned I encountered bad weather. I landed at Jackson's Mill Airport near Weston. To secure the airplane against strong winds, I hiked into Weston to purchase ropes and to get a room for the night.

Having purchased the rope, I then called a taxi and asked the driver to take me to the airport. His reply, "OK hop in." The driver stepped out of his car and made a telephone call. By this time it was dark. One mile out of town he stopped at a roadside inn—which is a polite word for beer joint. The driver blew his horn and out stepped a friend and he ordered his friend to get in he had some work for him. Twenty minutes passed and I was sure we were traveling in the wrong direction. It should have taken less than five minutes to get to the airport. It was time for some serious thoughts. Running through my mind was, "Is this robbery?" As I purchased the rope to tie the airplane, I also purchased a knife to cut the rope. This type knife in those days was known as a frog sticker. It was very long and slender with a sharp point—just in case I wanted to cut something besides rope. I opened the knife, unnoticed by the driver or his burly looking buddy. I was in the back seat, which gave me some advantage. If they started any monkey play, I just may go back alone.

As I put the question straight to them, I thought, "If I do not receive a proper answer, I will put the knife to the throat of the passenger and demand to be driven to the nearest police station." I asked, "Why have we traveled so long in the wrong direction." His answer surprised me, "Oh, I thought I had told you or asked you to go along." He further replied, "I am taking my buddy to my house to do some work." Soon he let his friend out and delivered me safely to the airport.

FAYETTE AIRPORT
February 7, 1974

Fog throughout the day had obscured the ridges surrounding Fayette Airport. As darkness came, it brought some clearing. The telephone rang. The caller, one of the girls working at Beckley Airport, asked for weather information. She explained that Beckley was 0-0 and one of their airplanes was overhead with no place to go. I reported Fayette as having three-fourths of a mile visibility, ceiling unknown, with a continuous improvement.

I assured the caller I would stand by the radio to assist the aircraft; 88X radioed Fayette Airport that it would try an approach. The pilot descended to five hundred ground level, making no visual contact with the ground. The pilot radioed again. I understood he would try Charleston, but a few minutes later I heard a plane. Looking out the window, I saw that he was crossing the airport about two hundred feet. I rushed to the microphone, placed outdoors to aid in visual reference to the airplane, and asked the pilot if he intended to make another approach. His reply was, "No. I am on the runway taxiing to the hangar."

The pilot left the airplane parked for the night. The next day there was fifteen inches of snow. The runway was in the process of having the snow removed from it when the pilot arrived for the airplane, a Cessna 310. The pilot was impatient to leave. As he started to lift off a door flew open, but he thought an engine had failed and he aborted the takeoff. Those watching ran to the field, thinking there would be a crash scene. But due to the heavy snow he brought the airplane to a safe stop and waited for more snow to be removed. He then departed safely.

AIR FORCE BASES

In 1939 and to the present time, circumstances have forced me to land at army or air force bases. Upon these few occasions, we are met with utmost courtesy, and only occasionally are we met by a big wheel who throws his weight around.

I apologize for the err, telling him that I started a little airport before he was born, and that more government airplanes have landed at the Fayette Airport than are based at that particular air base. And regardless of the way we are treated, we of Fayette Airport will render all the assistance, courtesy, and kindness to the pilots and crew that is in our power. What else can we do for those that have done so much for us?

Their attitude then is good and all is well.

DIXIE
1944

A young medical student flying a J-3 Cub became lost and landed in a small field parallel to the highway at Dixie, West Virginia.

Attempting a takeoff, the little Cub ran out of field. Pushing to abruptly put on his brakes, the Cub went up on its nose. He then called H. L. Sessler, Beckley-Mount Hope Airport, replacing the broken propeller. The pilot repeated his attempt to takeoff—but another busted propeller. It looked as though what this boy needed was a propeller factory.

Sessler returned to Beckley for another propeller and me. I agreed to fly the Cub out of the small field. There was not enough speed for takeoff and I aborted just short of the fence. Someone suggested flying off the road. There was a power line with a curve in the road. Once the throttle opened and full power was applied, I would be at a point of no return. It seemingly was the only way. If that did not work they would not need any more propellers, and they would be rid of me. When I reached the power line I had plenty of speed and over the top I went with room to spare.

FIRE IN THE WOODS

For a few seasons, we obtained a contract from the State Department of Natural Resources to fly forest patrol. The fire season is short—during the fall shortly after the leaves fall and early spring during a dry season before the leaves turn green.

The fires occur for various reasons—trains, cigarettes, campfires, burning brush, and arson. I have heard that lightning has set fires, but I know of no case.

The flying fire patrol is good, safe flying hands of the Department of Natural Resources. This has no resemblance of the Oak Wilt Blight Program, which has claimed the lives of so many young men. One wonders if our oak trees are worth the loss of lives from the Oak Wilt Program. It is in the hands of the West Virginia Department of Agriculture.

The Department of Natural Resources has a highly experienced pilot, Ash Kelly, who, through experience, knows what he is doing, keeping an eye on the program—safe for pilots, and keeping our forests green.

During the fire season when the visibility is very poor it is necessary for the airplanes to fly several counties when the fires are out of control. One airplane may spend several hours studying and directing the fire crews. Sometimes when the fires from the ground appear to be at their worst, it is only inside burning, (meaning there is a ring of fire but on the outer edge it has burned out). The pilot then watches for breakthroughs and radios the ground crew when one is spotted. The crew then rushes, stopping it before it becomes a big fire. The only unfair thing of this program is that the drafted fire fighters are only paid twenty-five cents per hour.

There were times when the chief forester would fly with us. One such occasion being on the summit of the mountain near Oceana. There is a fire tower by the name of Burning Rock, which is manned during fire season. This day the chief was riding observer. He spotted a fire on Alem Creek. By radio we contacted the spotter in Burning Rock Tower. The spotter replied, "There is no fire unless you set it." The chief quickly told him that he wanted to talk with him the first day it rained. This is a good program with money well spent.

This is the reason we ceased our service with the Department of Natural Resources. The federal came in complaining of a small radio aerial strapped on the gear of the airplane. It would be costly for a special weight and balance test to be done by an aircraft mechanic. We surrendered to the bureaucracy and their red tape and quit the program.

AIRPORT GHOST

Only one that has experienced such a night of terror can imagine what one goes through when he sees his first ghost.

Now we will turn the clock back many years to my teen-age days when I was working at a small airport near Mount Hope, West Virginia. It was a warm day in late February and while sitting in the airport office we heard an explosion. Running out we saw there were two pilots which had crashed and flames engulfed the plane killing the pilots. What a horrible sight for a young pilot as myself. Following, there was the unpleasant duty of removing the corpses which made a deep impression on all of us. When the job was completed, the manager called me in to volunteer my services to guard the wreckage through the damp February night, until the Civil Aviation Agency inspector could come and examine the wreckage.

Settling down for the night, alone in a dimly light gray painted room only one hundred and fifty feet from the wreckage, was no easy task. The air was cooling outside, creating a misty fog over the warm surface of the ground. The only cheerful thing was the little coal fire burning low in the heater. All went well for the first half of the night; then the dead still silence was broken. A sound came drifting through the open door. It was a whining, scratching, screeching sound. My better judgment told me it must be the wind blowing trees and bushes against the metal hangar or perhaps a door squeaking, but as reality came I realized there was no wind. I sat quietly listening for the sound coming through the misty blackness of the cold night.

It was not only my duty but curiosity and fear that forced me outside—stopping every three or four steps in the fog to listen to this horrible scratching, screeching sound. I had a six-cell flashlight and a .38 revolver, both being very little companionship. Perhaps the sound was only the fellows playing tricks on me, but at 2:30 in the morning out in the mist and fog this was hardly likely. If it could only be true. It seemed hours had passed as I moved inch by inch toward the sound. The sounds grew more weird and unhuman and at last through the mist I could see the burned fuselage. With my arm outstretched, holding my flashlight—to my horror I saw a ghost. I wished at this moment I could be a thousand miles away. Then the hopeful thought came to me that perhaps it was all a nightmare. Reality struck me again and there I was alone with the wreckage with two ghostly gray eyes staring at me from the bottom of the fuselage of the burned out plane where only a few hours ago were two corpses. All the fears that I had stored up in me for years had come to me at that moment. Either I had seen a ghost or I was in such a mental state that I needed help. Perhaps the day had begun to work on my nerves because I was chilling and sweating. I wanted to run but my legs were too weak to carry me, so I decided to draw the pistol. The light was still shining on the eyes and I could see them floating out of the old hull and move towards me. All the ghost stories I had ever heard passed through my mind, but it was little comfort to remember that Houdini once said there were no ghosts.

 My wonder grew—what demon had come to seek revenge? Could it be the devil that had come to claim lost souls? As it approached me I felt my legs grow weak and I fell in the mud. As I lay there in the mud I fought for full consciousness; I then decided to close my eyes and lay there in the mud and try to gain my strength to evaluate the situation. I felt a hot breath on my face; how much torment mentally can a man stand. Is this the way the wicked faced death? I must make one last attempt to free myself from this sinister creature which has conquered me. With my flashlight in my hand I opened my eyes. This creature hovered over me. I made one strike and to my surprise knocked it down. This was my chance. Now with the pistol in my hand I rolled over in the mud, shining the light on the creature. It was only a half-starved dog which had smelled the burnt flesh and was scratching in the ashes for food.

P.S. This story is true but intentionally overdramatized for interest.

CLARK FIELD
1948

W. Casto, school principal and weekend flight instructor, was working at his weekend profession—instruction at Clark Field.

While practicing takeoffs and landings with a student in a Piper PA-12 Super Cruiser, the pilot or student was seated in the front seat and the instructor in the wide back seat.

A sight-seeing cash passenger arrived at the airport. Casto, not wanting any of that green cash to get away, placed the passenger in the seat beside him.

This appeared to be legal except for one thing, Casto could not get to the dual controls, to aid the student if necessary. The student allowed the airplane to get out of control and hit a tree. The instructor was helpless. His passenger blocked the use of the controls. The three occupants were slightly injured.

HAVE A SEAT
1948

Assuming that you were with a very experienced pilot departing for your first airplane ride. Walter Summers accepted Judge Garvin's invitation for a sight-seeing trip. Getting into the airplane, Walter misunderstood the judge's instruction. The judge asked Walter to hold onto the steel bar bracings overhead of the back seat, and to go easy putting his weight on the canvas seat. He told him that it may be weak and would cause the seat to give way.

Twenty minutes had passed after takeoff to view the Bluestone Dam. The judge looked back to check on Walter and saw that his face showed much discomfort. Walter was holding firmly and tightly to the bars overhead, keeping all his weight from the seat. The judge asked him, "What is the matter,

Walter? Are your pants wet or something?" Walter replied, "No, Judge. You told me to be careful of my weight on the seat so I did not put any weight on it."

SMOOTH LANDING
1942

My girl friend finally agreed to an airplane ride. She was a lovely young lady and to me, it was very important to impress her with my skill. After a short flight, I knew that this must be the best landing I had ever made—and so it was. It was one of those with a lot of luck—so smoothly that you could hardly feel it had left the air and then it connected with the ground.

Wanda then said, "Does it always hit that hard?" This was the end of the courtship and that is why I am an old bachelor today.

SNAKE

There is lovely flat bottomland called Crumps Bottom along the most beautiful part of New River in Summers County. This is a paradise for a pilot who wishes to relax, swim, hunt or camp. It is located miles above Bluestone Dam. Occasionally, while flying Oak Wilt for the state of West Virginia, we would stop our clock and the observer and I would land, eat our lunch, and swim.

It was on such a flight that Jennings Erven and I landed and walked down the path that led to the river. Much to our surprise, we spotted a large diamondback black rattler with ten rattles and a button, singing his deadly song while laying coiled to strike. We gathered forked sticks, pinned him down, and put him in a lunch box. We then carefully laid him in the baggage compartment of the plane.

As we swam, we laughed, wondering what would happen if someone tried to steal our lunch box.

We continued the day flying a few more hours. Upon our

return home, the snake story stirred up much excitement. With the anticipation of seeing a live diamondback rattler, we reached for the lunch box only to find it, much to our surprise, empty.

With much concern, the plane was well searched, only to reveal nothing. We then figured that the snake had found a hole in the plane and not realizing the height, fell to his death.

The next morning the flights were continued as usual. As the plane was being checked for its daily flight, one of the gas boys commented, "Is that a battery cable on the floor?" It was no cable—it was the snake. Half of the snake was under the floorboard. A small ax was soon brought to the scene and the snake was divided. Giving it time to die, it was removed. One of the boys commented, "After flying with you for eight hours, why don't you solo the snake?"

SNOW AND MORE SNOW

There was an air ambulance call. The trip was to Morgantown to pick up a terminally ill patient. Flight Service reported en route weather as being perfect. No precipitation until night, and I arrived in Morgantown at twelve o'clock noon. I departed with the patient and attendant at 1:00 p.m. and the time in flight was to be one hour and twenty-five minutes. The weather was holding good. Morgantown Flight Service Station informed me as I departed that snow was beginning at Huntington. I replied, "I will be ahead of the snow. Fayette Airport is eighty miles east of Huntington." After thirty-five miles out, we were beginning to encounter light snow. I radioed Flight Service in Beckley and Charleston. Ceiling and visibility were good. The weather moved in on us fast. I called again to Beckley and Charleston weather. There was deteriorating one-hundred-foot ceiling and one-half mile visibility. Morgantown, behind me, was deteriorating also. I asked Flight Service for advice. They suggested I go to Parkersburg. I rejected. Morgantown was bad, Huntington and Beckley below minimums, Parkersburg would soon be the same. They soon called and confirmed that Parkersburg weather had deteriorated.

I asked where there would be good weather. They said Elkins. Again I rejected, believing the storm was moving too fast and the mountains were too high.

I asked about Roanoke. This, they said, was very good weather. It seemed to be my only chance. Radar soon picked me up and monitored me in the direction of Roanoke. They informed me I was twelve miles out and was clearing me to descend to 3,500 feet. There was no visual contact with the ground. Approach control advised me that I was now four miles out.

Then there came a clear order, "73046, climb immediately on present heading to 5,000 feet." I soon replied, "Roanoke approach level at 5,000." Roanoke then said, "We cannot get you in. Visibility is one-eighth mile." My reply, "What is Lynchburg weather?" Lynchburg was good with ceiling 7,000 and visibility six miles. They requested I hold my heading and I would run out of the snow in six miles.

Roanoke turned me over to Lynchburg approach. Lynchburg asked me what type approach I wished to make. I told them I thought their weather was good. Their reply, "Yes, but it is deteriorating rapidly." I requested to descend to an en route altitude. As I descended there was the beautiful Lynchburg Airport in front of me. The tower cleared me to land.

I immediately called an ambulance. The snow had not started yet. Within thirty minutes there was four inches on the ground.

Seven hours later we arrived at Oak Hill Hospital. I owed the passengers a debt of gratitude for their patience with me. They offered to pay, but there was no way I could accept money from these fine people, they being the Reverend Mr. and Mrs. Meadows.

SNOW GO
1951

February brought with it snow, paralyzing our roads and all the airports in southern West Virginia. All travel was near a standstill. The State Road Commission was busy trying to open the main highways. The snow stopped, but the roads were not yet suitable for distant travel.

The hospital called asking if I would take an emergency trip to Richmond, Virginia. I told them, "I cannot today. By tomorrow the State Road will surely clear the runway for such a trip." I called the local highway department, but they were not cooperative. Their answer was not definite, only a put-off.

I called the hospital telling them the trip could not be made. The hospital called back and asked me to call the governor, the governor being our own Okey L. Patteson.

After placing the call, I asked the secretary if the governor was in. Her reply, without question, was, "Yes, do you wish to speak with him?" As I was telling Governor Patteson the story, I asked him if he desired to check with the hospital. The governor said, "I have heard all I need to. If the State has not cleared your airport in twenty minutes, call me. I will remain here for your call." Before the twenty minutes was up, it looked as though all the equipment the state owned had arrived. I then called Governor Patteson to thank him, and I was amazed when the governor answered the phone himself. The trip was made successfully, thanks to a good governor.

SIX DEGREES BELOW AND GETTING COLDER

Dave Burgess and I, late one November evening, journeyed to Cuyahoga County Airport, Cleveland, Ohio, for the purpose of transporting a deceased. Arriving at Cleveland, the temperature was six degrees below, so we decided to wait until the next morning to return. We tied the airplane, securing it against hard wind. The attendants assured us that they would gas the airplane before closing.

Reaching our motel we found the room very cold, so we called the desk. They came and confirmed that the room was cold.

We were up and ready to depart at daybreak. Finally someone offered to drive us the two miles for ten dollars, only to find the flying service was not open. We hovered in the aircraft to get out of the bitter cold and wind only to discover they had failed to gas the airplane as they had promised. When they arrived and opened, two hours later, they could not get their gas trucks started. By twelve noon their trucks started and they refueled us, but the aircraft would not start. They informed us that for ten dollars they would heat the airplane, but while bringing the heat to the airplane, their fuel ran out of the heater. To town they went for a fresh supply, causing a one hour delay, after which they heated and started the airplane.

I called the funeral home to bring the deceased out. He first took it to the wrong airport, then to the correct airport, but the wrong flying service. By 2:30 p.m. we were ready to fly, receiving our taxiing instructions before we reached the runway; the field visibility went to one-half mile which was under another set of rules, so we were not legal to depart.

It seemingly was not meant for us to leave that day. The snow squall did pass, clearing the weather without further difficulty until reaching home base, finding very deep snow had fallen. We landed with much concern but without trouble.

HURRICANE CAMILLE
August 1970

A call came to make a quick trip to Fort Myers, Florida. Larry Hess accompanied me. At Winston-Salem, my usual route was fogged in. This was as much as I knew of the weather, so I decided to obtain the weather briefing from the FAA while in flight. The answer sounded like a nightmare. There was fog in North Carolina and tornado warnings in Georgia. I asked, "What if I fly west to go around toward Chattanooga, Tennessee?" Their reply was, "Towering cumulus thunderheads with tops to 40,000 feet." This looked like the only possible way, so west we went, circumnavigating many thunderstorms. I finally called Chattanooga radio asking the weather south of Georgia and Florida. They said, "Don't you know hurricane Camille is due to come ashore tonight with winds 125 mph?" With luck, we made it the next day. Hurricane Camille changed its course and headed for Texas. We were able to fly home without further trouble.

After Hurricane Camille hit Texas, it changed its course coming East, causing loss of lives in West Virginia and along the James River in Virginia. I again missed another storm by a few hours. I was driving the scouts to Washington, D.C. We were routed around Rainelle because of high water. The roads in the Covington, Virginia, area in places were nearly impassable with rockslides. We were thankful that we had barely missed the bad weather so many times.

INSURANCE
The Summer of 1958

In the black and dark hours of the twilight evening, the phone rang. "Frank, this is Marvin Dodge from Lewisburg." Someway, with his smooth, fair speech, he had me believing we were old friends. He wanted me to join him at the local Hotel Hill to talk over old times. I agreed, not realizing that I was about to be taken, swindled, and cheated.

He said that he had not seen my airport for some time and suggested that we visit it. As an ox goes to the slaughter, we headed straightway for the airport.

Marvin was an insurance man and a lay preacher. He asked me about my insurance and I explained that I carried only liability to protect my customers. With much flattery, I yielded to his salesmanship. He left town with a large bag of my money, being five hundred dollars, and was to return at an appointed time with an insurance policy. He assured me that the policy was in force immediately for everything except my dog.

Ninety days passed and during that time I called him several times. But still no policy. He explained that it was customary for the insurance agent to keep the policy in a good safe place and said that it was obvious that I did not trust him so he would fly the policy over immediately. I replied, "You may be right. Prove me wrong, Marvin, by bringing the policy. I do not trust you." He then said, "I'm on my way."

He arrived shortly, and I told him that I was sorry that we could not have a brass band out to greet him. Marvin pointed to his briefcase, "It's in here." Searching his case, his plane and his pockets, Marvin then stated, "It's on my desk. I'll have to go back after it." "No," I said, "just mail it by return mail."

Marvin then asked if I would take him into town because he wanted to get a small check cashed. I was waiting patiently at the bank door when Marvin came out saying, "They want some identification. You will do." With loving force, he gently dragged me in, turned the check over for me to endorse, and said, "It is only for five." I replied, "Why didn't you tell me; I could have cashed it."

A few days passed with still no insurance, when the bank called that the check had bounced. They said, "We will have to take it from your account." I replied, "OK, what is five bucks." Then they told me that it was not five bucks, but five hundred. I had been bitten by the same snake twice. That five had two zeros in front of it and now he had gotten me for a thousand. I always thought that while he was looking up he was looking for the Lord, but then I realized that he was probably watching for a flying sheriff. Now I was wiser, much poorer and very mad.

I called him and told him that I was getting a warrant for him and he said he did not understand me, that he thought we were friends. He then said that he would fly over and explain. He came.

I then found two other airports that he had sold insurance to but had not delivered the policy. I called lawyer Renick and Marvin thought he was the sheriff. A telegram then came from an insurance company stating that we had no insurance with them. (Marvin had tried at that late date to insure us.)

Marvin said, "I will do anything to keep from having a scene." "OK," he was told, "Sign some papers and if you don't pay up within thirty days, the Bellanca Airplane is ours." We typed the papers and held them.

I then called the West Virginia Insurance Commissioner. He told me that I must be mistaken, that the Dodges were fine people. I told him, "You must be part of the swindle." Mr. Renick took the phone from me and told the commissioner that I was correct; that Marvin was kiting checks and not delivering policies. The commissioner then said that he would start driving that way immediately. The commissioner investigated the case, revoked Marvin's insurance license and Marvin was put on five-year probation.

After thirty days, as we watched the sun go down, in came Marvin. He paid up. I had let the air out of the tires of his airplane, thinking that he may return while I was gone, and he thanked me for doing that and then departed.

We later learned the reason for his downfall. He married a New York girl and had told her that he was rich. She believed him and started spending.

MORRIS BURNSTIN

When skill and luck meet face to face, such was the time Morris Burnstin had his first forced landing.

He had been practicing stalls high over Fayette Airport when the engine sputtered and gave up the ghost. Probably he had failed to add a little power as he was practicing, as this was common on the 65 hp engines in the J-3 Cubs. Nevertheless, when it came time to head home, he circled the airport twice with skill and precision. When he turned on final, he was too high to land. With no power to go around, all that appeared to be before him was the woods.

We ran to our cars in order to be at the wrecking as soon as possible. Crossing the threshold, his height appeared to be at least three hundred feet. In the J-3 Cub, with a ten to one glide ratio, that would mean he would touch down three thousand feet from this point. The airport being only a little over two thousand feet, this would put him well into the woods.

We all rushed to the car, and after climbing aboard, someone said, "Who has the key?" It was not to be found. The airport, to say the least, has a slight roll; or wording it better, the plane was out of sight over the hill or mountain which the runway extended, or had gone deep into the woods. We started our two thousand foot dash, finding the airplane had finished its roll at the total end of the runway.

THE UNINVITED GUEST AND SQUADRON THREE

In the year 1949 I was asked to be an adviser of Air Explore Boy Scouts of America (BSA). The years that followed were full of enthusiasm and excitement with never a dull moment.

There were times when Squadron Three was the best, and no doubt, sometimes they were the worst. One evening a week for twenty years, Squadron Three met at Fayette Airport. One book would not contain all the good projects and activities. But, these airport stories would not be complete without one amusing, naughty activity of these restless teen-agers.

A few of these boys were camping on Fayette Airport each weekend (these were the days before the local drive-in theater). And with all the efforts put forth, there was no way to stop the motorist from opening the driveway gate, and driving in to use the airport as a lover's lane.

To play jokes on these uninvited guests was the highlight of the evening. A car would pull up and park, then minutes later the smallest of the boys would go to the car, tap on the window and ask the driver the time. The young boy would then accuse the driver of lying about the time, intentionally provoking the driver to the point of desperation. He would then threaten the youth and the young man would dare the driver to step from the car. The driver, in no way allowing his girl friend to think he was a coward, would step out. The youth would retreat and the first line of defense, twelve to fifteen, would come out of the darkness with their ball bats asking, "Can we help you?" One of the boys always had a bucket of water to throw on the man to cool him off and dripping wet, he would look up and say, "Thanks a lot; I needed that." He was sometimes told that he was the third man that week that had been in lover's lane with that same woman. He was told, "If you come back we will call her husband."

During his recovery he would discover his car had been jacked up and put on pop crates with tin cans tied to the car.
Our uninvited guest, once departed, never returned.

THE CESSNA 120
1946

It was a hot day in August. Carlton Clark was a well-known pilot in this section. Clark landed at Fayette Airport to try to sell a new Cessna 120. Fayette Airport was, in those days, only 1,100 feet long. I told him we had heard its performance was not equal to that of a J-3 Cub for short field takeoffs. He assured me it was and asked me to go for a ride. We were interested but first he had to prove it would go out of the airport with one person.

He agreed, but apparently he did not judge the heat, the rough field, and the altitude, which all add to a slow takeoff.

On his takeoff run he used all the field and went out of sight. He had found the little hidden ravine that slowly leads to New River Canyon. After two and one-half minutes we saw him climb out of New River Canyon, safely homeward bound never to return to Fayette Airport.

Mr. Buzzey Zoom
Furnace Room 13
Hades 666

Dear Mr. Zoom:

We know you were the best pilot that has ever flown. Most people were envious of your ability to fly under bridges, power lines and rolling your wheels on rooftops. Jealousy is the reason why you were fined. We do not believe you were drinking too much. You always held your liquor well. And even if you did take the plane without permission, they should not call that stealing.

We realize that you enjoyed those fast-climbing turn pull-ups close to crows, and enjoyed the expression of their fear and panic as you passed low over their heads, blowing off their wigs and scaring their kids and cats.

The flowers at your funeral were beautiful. The fear of your neighbors seems to have turned to a quiet peace. We will remember you every time we pass your grave. We wonder how such a skilled pilot hit that bridge. They certainly should have built them higher from the ground.

OMEGA

P.S. I am asked daily who will be the first to fly under the New River Bridge when completed. I know only that it will not be me. I do hope I will be around to report them and see them fined or jailed.

THE COST OF BUZZING
November 14, 1956

The only reason a story of such sadness and carelessness is added to this collection is so that it may serve as a warning to all that have knowledge of it.

On Monday, November 14, 1956, 4:10 p.m., Spears and Sweany departed Beckley-Mount Hope Airport to deliver some sandpaper in an airdrop to a garage at Coal City, West Virginia. Arriving over the town and nearing the garage, Spears turned the controls over to Sweaney so that he could concentrate on the package drop.

The plane made several passes, then struck a power line, smashing to the earth causing the worst of tragedies. It ripped off the right arm of Herbert Justus, a ten-year-old boy, who was playing in his own yard. The child's death came only a few hours later in a local hospital. Mrs. Justus, weeping, cried, "You will never know what this has done to me."

It was learned later that the cap from the plane's gas tank was found near the pump at the airport. The fuel may have been sucked out of the gas tank by the wind while the plane was in the air.

Both occupants were licensed pilots and the airplane was a Luscombe.

This is one thing that cannot happen to you if you do not buzz or fly low.

PROP WASH

Prop wash is the term used to explain the blast of air created by the propeller.

In 1949, John M. Frazier was polishing on his father's new airplane. John was youthful, innocent, and not aware of the pranksters at the airport. Harold Stewart and Frank Vest convinced John that it was important he have some prop wash.

Off flying to Pence Springs, John went in search of prop wash. Jim Tolley, sympathetic of his need, gave him a bucket full of dirty motor oil on his return. John vigorously applied the dirty oil to the propeller before realizing it was used oil.

PROPELLERS

Has it ever occurred to those who fly at Fayette Airport why there is such rigid rules for shutting down the engine and stopping the propeller while loading and unloading passengers?

It was Sunday afternoon and we were hauling sight-seeing passengers. As passengers were being unloaded and directed to the direction of travel, a small boy crawled under the door and strut and was very near the propeller. When Joe Martinez rushed up and grabbed the small boy by the collar, he was only inches from the whirling blade. Pulling him back, Joe said to him, "Where do you think you are going?"

Thanks, Joe, for saving a life and helping maintain our safety record.

This was the same Joe who, on his first solo flight, disappeared and flew over his house to buzz to let his family know he had soloed.

AIRPLANE PROP KILLS BUSINESSMAN
January 7, 1974

A businessman apparently had his mind on other matters when he walked into the whirling propeller of a single-engine aircraft after he had landed it last night at the Wood County Airport.

Robert C. Kylberg was killed instantly after stepping from the aircraft en route to moving his car parked in his aircraft's tie-down area, sheriff's Lieutenant James George said.

"He apparently just had one of those little mental lapses we all have at times," Lieutenant George said.

The accident occurred at 7:10 p.m.

THE BIG CHASE
Princeton 1941

W. R. Capidonia, a young man from Welch, West Virginia, attempted to start his aircraft by hand, although he did not have an assistant at the controls, and did not have the wheels blocked or tied. After pulling the propeller through several times, it started.

The aircraft first crept forward. Capidonia would move forward and so would the airplane. Vibrations advanced the throttle causing the airplane to move faster, chasing Capidonia across the airport. The airplane caught up with him, clipping the tail of his overcoat, which he skinned out of.

He reached the fence before the airplane. Capidonia, during the race, had enough speed to jump the fence, but the aircraft did not. The fence stopped the airplane with only slight damage to the aircraft and fence. Major damage was done to the overcoat.

WHY THE SIGHT-SEEING TRIP IN 1942?

A well-dressed man asked for an airplane ride. This turned out to be most unusual. He requested plenty of gas in the airplane, saying, "I am paying the bill. I will tell you where to fly and when to come back."

He wished to fly to the Beckley water tank and circle. We circled a few times. I then said, "Where to now, Mr." "Just keep circling," he replied. Two hours and forty-five minutes later, I returned for gas. He paid, appearing tired and disappointed.

The next day he returned. He asked if we could try again. My question was, "Try what?" He replied, "Oh, just take me where you took me yesterday. If you get tired of circling to the left, you can circle to the right." The afternoon passed slowly after three hours and seventy-five circles. We landed and he said that he was disappointed. I asked if it was my flying. He didn't reply.

On the third day, the same man came and requested the same flight to be taken. This time I demanded to know what was going on. My thoughts were that he was trying to gain nerve to jump out and commit suicide. Putting the question straight, I said, "What is this about?" He explained that his wife was running around with another man and he was trying to catch her from the air.

She must be a catbird.

FUN AND FOOLISHNESS

Hillard Stone was one of the most dignified, well-dressed persons we had ever had at the airport. We sometimes called him doctor.

A young man approached Stone, believing him to be a doctor, and asked him for an aviation physical.

Stone told him to go into the office and disrobe. He asked, "Everything, Doc?" Stone replied, "Yes. Call when you are ready. Don't fool around. I want you looking like a baby Blue Jay when I come in."

When the youth called, "Come in, Doc," Stone had already gone home. Six of the airport crowd rushed in, pretending to be surprised, asking the youth, "Where are your clothes?"

RABORN COOK
1946

It had been three years since Raborn Cook had used his talents at Beckley-Mount Hope Airport. Raborn had just returned from service.

Early one rainy morning, Sessler asked him to make a charter trip to Florida. The flight was to be made in a Cessna VC78. This airplane was covered with a fine grade of fabric, unlike the metal aircraft of today.

While Raborn was loading luggage, the lady standing beside him, to be one of his passengers, closed her umbrella, raised it high, and stabbed the umbrella through the wing of the airplane. Turning to her friend, she said, "I told you this was not metal. I ain't getting in no airplane made of cloth."

THANKSGIVING DAY
1947

By telephone we received a message that there had been someone parachuting from an airplane over Oak Hill.

The usual airport group was gathered in the trailer office. Among the group was John Wilson, later to be the West Virginia aeronautical director. John was anxious to start a search. He took to the air immediately. Someone called and identified the object as a roll of toilet tissue that had been thrown from an airplane.

Someone suggested a joke to be played on John. Quickly we dressed Howard Chambers, a sixteen-year-old youth, into a flying suit and helmet, carrying an open parachute. Howard was sent over the hill out of sight not to return until after John had landed. When he landed, John returned to join us in the office. Mrs. Jean Wise Crouse, watching casually from the trailer office, pretended to be surprised, seeing someone with an unfolded parachute under his arm. John Myles, now Dr. Myles, was in on the hoax. He ran with Wilson to assist the stranded pilot. We all followed. Wilson asked Howard what had happened? What was the plane type you lost. Howard did not answer. Wilson said, "The poor man is in a state of shock. Get a doctor quick." When Chambers reached the office, Wilson asked, "Can we call your commanding officer? Is there anything we can do to help?" Chambers answered, "Yes, help me out of this suit and garb and give it back to Frank."

GAMBLING IN 1945

The Veteran's Flight Program was in its beginning. The GI's were coming home.

With the excitement of all the years behind, they were constantly looking for action. While waiting for an available airplane, someone brought out a deck of cards. Anyone who had a few pennies could get in the friendly little game. It seemed to be a game of skill. The limit of your loss was one dollar then you must quit. The house took nothing. Weeks passed and the penny games had lost their thrill. So then it was dollars, followed by twenties. The games lasted sometimes all night, with as much as one thousand dollars on the table at one time. It did not take long for the professionals to arrive. So much money changed hands you would think you were in a bank. Flying business dropped off. The excitement of the card game was too much to leave to fly. The reputation of the airport was at stake. I came to a quiet and quick realization. I asked the boys to make this their last hand. The game is over.

The sun did not set on the next day until the word was out. Frank has cleaned house. The flyers were back. To my knowledge there has not been a card turned at Fayette Airport since.

SOUND PLANE
1948

As May and primary election time arrives, the little loud Cub with the powerful public address system was one type of advertisement that could not be shut off. If you were in a soundproof building, the sound probably came through the walls.

Before I purchased this airplane it had been advertised as the second loudest public address system ever put together. In previous elections it had brought much comment, good and bad, both in West Virginia and Kentucky.

Someone introduced a bill in the state legislature calling it a nuisance and asked that it be quietened and banned from the sky. M. R. Renick, an attorney, represented me before the legislative committee.

When we reached the committee room, we found Mrs. Nell Walker, a woman legislator. Mrs. Walker was known as the first lady of Fayette County. She was highly respected with plenty of experience in the legislature.

We knew the chances of the survival for the little loud Cub—final ruling rested on the decision of Mrs. Walker. One hour passed and all of the committee had had their say. Some were for and some were against. All had spoken except Mrs. Walker. Mrs. Walker, with all evidence in, rose to the occasion in our defense. She told how the little loud Cub tried to help apprehend the robbers of the Winona Bank and how the airplane was over the Richwood flood to offer assistance. She stated, "We are proud to have it in Fayette County." She would fight for the bill in the committee. Then, if necessary, on the floor of the house, that this method of communication should not be silenced. The bill was dead.

FREEZING ON CONTROLS

As I was learning to fly there was much discussion of students freezing on controls and tightening their muscles so tight that it was necessary to pull them from the controls by force. If this has happened, I cannot say. It has not happened to one of my students.

However, I experienced one case of a student on a final approach for a landing who pushed the controls all the way forward very rapidly. I grabbed the controls and pulled it out just short of the trees.

One other student responded to every request with a robot hypnotic action as though he were in a trance. Upon landing he declared that he had not been up and refused to pay for the lesson.

One of the Parkersburg instructors told the story in 1964 of a student becoming annoyed with his instruction and trying to throw him out of the airplane. It was hard for me to understand why a student would allow himself to lose control of his emotions until I flew with this man—then I understood.

DECEITFUL

A young man, with uncertainty in his voice, approached me. His story was that he had some flying time, but very little. His approach at first seemed sincere. He continued that he needed help and fast. My answer was, "You came to the right place. What is your experience?" He replied that a friend had told him that I could be trusted. His story was frightening. He told me he had a commercial license and as the story continued, he said he had never soloed an airplane, but was a legal A&P mechanic. He had obtained a commercial license but it was not legal. I said, "This is interesting. Tell me more."

He said that he and a friend roomed together in Florida. Both were going to a large aviation school. He was to be an aviation mechanic and his friend was studying to be a commercial pilot. They had duplicated each other's papers and had taken each other's test from different examiners. He said, "Now you know. I need to be started all over again. Get my flying by myself. I have a flying job waiting. I will be careful while gaining experience."

My answer was quick. It was no. I wonder how often this has been pulled before. What number of accidents have occurred due to pilots without training.

It is rumored that an airport manager in southern West Virginia has a commercial license and has never soloed an airplane. I have been unable to find anyone who has ever seen him fly alone.

TO FLY WHAT?

It was a proud day for Douglas Epperly. He had received his private pilot's license. In those days it was not easy—spins, stalls, precision landings, and all the instructor and examiner could throw at him.

That big day was here for Doug had completed his hours and training. The flight check was completed and approved and all papers properly signed, except one. The examiner, for a joke, in the bottom copy of his temporary certificate, had typed, "License to fly a kite during the month of march."

Doug departed proudly with his license. Walter Summers, a friend of Doug's, arrived on the scene by calling Doug on the telephone and asking him to read what the license said. Douglas returned to the airport immediately to exchange for the real thing.

EMERGENCY MESSAGE

The telephone rang and the first words were, "Where is Frank? Get him to the phone immediately." The one answering said, "He does not like to be stopped when he is flying. He is with a student." This was my sister calling, "This may be life or death. I have an urgent message for him. Quick! As soon as possible get him to the phone. This is very important."

Three other airplanes and myself were practicing landings with students. Upon landing, I was motioned in to the gas pump. I terminated the lesson with the student I was then with.

Walking to the phone, I directed another student to gas the

airplane for one of the many Roberts we had flying with us. I told him to take the airplane and fly alone, while I called my sister to see what the trouble was.

When I called her, the phone only jingled once. "Hello. Is this you, Frank? Where is Robert? When he left town he told several people he was going to commit suicide with one of your airplanes. He says he will fly it into Oak Hill water tank. He must be kidding." I told her I would have to stop him immediately because the airplane was being gased for him.

By the time I reached the area, he had already started taxiing to takeoff. I ran to my car to head him off. Before I reached my car he was at the end of the runway ready to fly, except for two airplanes on final approach for landing.

By the time I finally reached there, the other airplanes had landed. The runway was clear and he was pulling in position for takeoff. As a desperate method to stop him I pulled the car up on the runway in front of the moving airplane, believing if he did not stop the impact against the car at this speed would be slight and would only damage the airplane and car. He stopped inches short of the car. He admitted his intentions and I suggested he do his flying with my competition.

FAYETTE AIRPORT
February 9, 1973

A gloom of sadness has reached Fayette Airport. Pete Puckett, the flyer among flyers, is dead.

All agreed among men that there has been none with such a love for flying. Where else would it have been fitting for Pete's passing than under the wing of an airplane.

After a full day of flying, he landed, stepped from his plane, and went to sleep.

All that pass his grave will say, "Here lies a flyer; here lies a man."

When duty called, he was there.

HANDBILLS

It was early May, around the time of primary election, and all the Fayette Airport crowd was excited. One of the candidates running for sheriff of Fayette County was a friend named Roy Swanigan.

Roy was a double amputee, losing his legs in a mining accident at Summerlee. He overcame his handicap by providing well for his family. For his pleasure and travel he learned to fly, becoming a private pilot with great skill.

We, of the Fayette Airport crowd, decided to help Roy by putting out one-half million leaflets asking the voters to support Roy for sheriff. But, as usual, there were a few thorns among the roses, one being the litterbug law. There was yet another way to view the situation. If the court fined us for litterbugging, it may very well be the publicity we needed. Because we remembered that in a previous election there had been a man from Ansted elected to the house of delegates while serving a jail term.

With these leaflets we intended to snow under a few towns. Mr. Renick found out our plan and pleaded for Fayetteville to be spared. His pleas were as though we were going to drop a bomb, being so urgent that we granted Rusty his wishes.

We well remember years before when several airplane loads of leaflets were dropped on Mount Hope. It had been a surprise attack election day dawn hour. Bernard Rock was running for mayor on a write-in ticket. This put the message across and Rock was elected.

Now, the problem at hand was what to do with the one-half million leaflets. While on cross-country, I could safely distribute several hundred. It would be easy considering you were leaving the state not to return for several hours. Time was growing short, when B. F. "Pete" Puckett, Jr., arrived. Pete well knew that if the handbills were not distributed at a low altitude the wind would carry them into the woods. This would be good for the hunters the following fall, because they would not have to carry any paper with them.

The zero hour had come; Pete was off, (that is to say, in the airplane). A long wait of five minutes had passed before the action started. Then the telephone started ringing. The man calling was, according to the *Daily Mail*, the second richest man in West Virginia. He was very concerned, stating that his farm had been ruined, and wanted to know what I would do about it. I suggested he have the butler pick up the papers, then I would come by and pick them up and try again.

There was no waiting for the next call—"There is someone going wild in one of those flying machines, trying to hand people some papers. He blowed the old ladies clothes clean off her clothesline. It's your plane, cause it's yellow. I seen you flying it out younder last Sunday."

Next call—"Hello, I'm at Gatewood. Some low flying X ! X ! X ! hit my horse on the rump with a bundle of papers. He bolted through the fence and we ain't been able to find hide nor hair of the critter since. I was a gonna vote for Roy cause he hailed from this part of the sticks, but not now. And somebody is gonna get sued."

On the next call the operator said long distance, so I knew Peter was moving on and getting attention—"There is an airplane flying too low here. I am at Meadow Bridge." He asked if it was my airplane and I told him I did not know. I asked him if he got the number, and he replied, "I'll give you more than that; I'll give you the tail wheel he knocked off on my chimney. You come over after it if you want it and I have something else for you." I suppose he has the tail wheel yet.

It was easy to tell that Pete was on his way back. The calls clearly traced his route. We hoped he would somehow gain a little altitude and make it before some kids downed him with rocks.

A call came from a lady from Scarbro. She said one of my boys was flying too low. She said he was smoking in the airplane. I told her we did not permit smoking in the Cubs.

The hundreds of pounds of paper were gone and Pete was homeward bound on final about to land. There was a call for him. Some lady said he had scared her cat causing the cat to give birth to her kittens prematurely.

Look who is driving in—State Police Trooper Lilly. He said,

"Thomas, I am taking you in. You scared a man's turkeys nearly to death." I asked, "What if you have the wrong man? You know what will happen if this is false arrest? What if the airplane came from somewhere other than Fayette Airport?" The plane tail assembly was covered with leaflets. Trooper Lilly departed empty handed.

From then on, we were careful about the remainder of the leaflets. We waited until there was no wind, but foggy. I believe this was the big factor leading to the defeat of Roy for sherfiff. By the time of another election, the voters had forgotten. He was elected to the house of delegates with ease.

GI BILL

In our search to enroll new students in our flying school, a letter was sent to Col. Shirley Donnelly, a local Baptist minister. We explained that flying with us were doctors, lawyers, merchants, and chiefs, and that we would like to enroll him.

He replied in his newspaper column. The saying is rich man, poor man, beggar man, thief. Which did we consider him?

BAD CHECKS

Appearing among us was a man with the dignity of an earl or count; a true gentleman. Dressed like a little English lord, it was easy to see he himself was impressed with his own importance.

He was in a hurry to learn to fly. We hillbillies were almost embarrassed to be with such an elegant gentleman. After starting his lessons, several weeks had passed, and through much embarrassment, it was time to ask for some money. He said, "Certainly, sir, I will give you a check." He did. But, the bank seemingly did not recognize his importance. To make a long story short, D. M.'s check bounced.

Such checks are thrown into the back end of a fireproof vault and guarded night and day by my faithful dachshund until the stupid sonzey makes the mistake of coming our way again—this being the day of reckoning.

Such was the day, ten years later, when D. M. called for a private pilot flight check. The man did a wonderful job of flying, meaning neither of us were hurt during the flight check. All papers were carefully signed. Now he was a private pilot, he thought. I spun the vault open and put the papers in it. He asked if there was something I was supposed to give him. I said, "Yes, there is something I have wanted to give you for a long time." He was given the bad checks and he nearly passed out. He was weak and pale when I reached for the oxygen bottle to administer first aid. He rallied, saying, "Don't hit me with that."

Pulling out his wallet, he emptied it, and with a clear conscience and sad heart, he left all those nice green pictures of the father of his country. He departed a private pilot and we were very proud of him.

JENNINGS WAGON

There were reports each Sunday of a Cub airplane landing in a hayfield south of Oak Hill. After a dozen Sundays, the mystery was solved.

The pilot walked into Fayette Airport, because he had damaged the propeller of his plane on a stump. The Reverend Mr. Wagon introduced himself as a dignified Holiness preacher and convinced me of his honesty. This led me to selling him a propeller—on credit, and to prevent further damage to his airplane, I invited him to make his landings at Fayette Airport.

The following Sunday he arrived and was seeking road transportation to the radio station. I loaned him the airport limousine—which was an old pickup truck, not realizing that he was going to try to make a career with the use of my truck. After using it for eighteen weeks, he never did put a drop of gas in it, but always used it as if it were his own.

Occasionally I would ask him for payment for the propeller. Rejecting my request, he informed me he could, but would not, pay on the Sabbath Day. He told me I was a sinner for asking him.

I finally told him I was going to take the propeller for nonpayment. He did not believe me until he found it gone. When he returned from the friendly radio station, he was very unhappy. He sounded more like a sailor than a preacher. With him was one of the sisters of his church. He was embarrassed and was going to the police.

The next Sunday he did not show up. The telephone started ringing. My friends were calling, asking me to turn on my radio. I was being preached against. As I listened to the Word being preached, I learned that if I was to believe him, I would have to believe that I was Satan himself.

Weeks later he returned with a bundle of money, begging me to sell him the propeller. I agreed and took the money.

He told me to put the propeller on his airplane. I informed him I was not an aircraft mechanic. I could legally dismantle, but not put back. So I loaned him my tools to put the propeller on his airplane. Now, where are my tools?

PAUL KING
1947

Paul King was an early enthusiast of flying and an excellent mechanic who owned a J-2 Taylor Cub. He departed, flying from Fayette Airport, early one morning while the fog yet covered the hills. Many anxious hours passed before the telephone rang. Paul had become lost, landing near Whitesville. We went to Whitesville. There, a farmer gave us a complete, vivid, detailed description. He said, "Paul circled the corn field, but his approach was too high and he touched down in the far end, ran clean through the barb-wire fence, jumped that there creek over younder, ran through the blackberry patch and hit the chicken coop and killed the rooster that Maw is cooking now. Then he let that airplane get out of control and hit the outhouse." There were no injuries to Paul.

FELL ON ME

As a student was practicing landings in a cub trainer at Fayette Airport, on his final approach as he was nearing the ground when he should have been slowing the airplane down, he dropped the nose, gaining speed. He then abruptly raised the nose and the plane climbed until it lost all of its flying speed. This created a complete power off stall several feet from the ground, causing the plane to fall hard damaging the landing gear. The repairs for Cubs were minor if you had the parts, and we did. One person could replace the gear in one hour. We pulled a lift to the place of damage to raise the airplane, the lift being on wheels.

I raised the airplane very high with the lift, four or five feet, for room to stand or sit and work under the plane.

Prying the first half of the landing gear came easy. The second half was a different story. I pried with a very heavy bar to free the jammed pieces.

At that moment trouble began. A strong puff of wind came. I saw by the shadow of the plane that it was falling. The lift was turning over. I dived for safety, but it was too late. I was pinned to the ground under the airplane. I was pinned near the position of the airplane known as the fire wall, in front of the broken gear and behind the exhaust stacks, and inches on either side would have crushed me or run sharp metal through my back.

I was alone, trying desperately to free myself. I could hardly breathe. I thought, "What a stupid, clumsy thing to have raised the airplane with the lift, so high to allow it to fall on me." As I found myself weaker, I dug my fingernails into the ground, knowing with limited breathing it would not be long before I would be unconscious. Inch by inch, slowly I pulled myself to the point where there was no weight bearing above my chest. I could breathe again. Gasping for air I could feel my lungs start to react. After resting for a short time, my strength came back. Struggling and tugging, I was free from beneath the airplane, only to discover the dog was caught by the back section of the airplane. If I lifted the nose of the plane he would be free. This was too much weight for me. If I lifted the tail he would be crushed. I called by telephone for Roy Swanigan, a neighbor, to come to the rescue of my dog. Neither the dog nor I suffered any injuries. The airplane flew again late that afternoon.

Flying
History

JACK BROWN
1918-1975

Jack Brown's contribution to aviation and his fellowman is unquestioned. One short story such as this cannot do Jack justice. It would take an entire book.

Jack, a native of West Virginia, started flying at the age of twelve, at Charleston. Transferring to Beckley-Mount Hope Airport for instruction from H. L. Sessler. There he received his instructor's rating as his love for flying flourished. He became a World War II pilot for the navy, flying in the Pacific. He later instructed at Bartow Air Base. For a short period he was a jet test pilot.

After the war he returned home to Gauley Bridge, West Virginia. Finding himself restless for flying, he returned to the land of sand and sunshine where he had some of his early navy flight training. There at Winter Haven he established a sea plane base. He achieved much success and great prominence.

Jack's field of endeaver expanded, including flying for motion pictures. Jack flew a Tiger Moth, and a World War vintage biplane for the movie, *Nothing by Chance,* a motion picture documentary on barnstorming. It was filmed in Colorado, the Midwest, and Winter Haven.

In 1974, sources say Brown trained 350 pilots for their federal license.

When Jack left Winter Haven, November 13, 1975, he borrowed a sleeping bag saying he intended to camp out while flying in the mountains.

A move reminiscent of his barnstorming days. He was a barnstormer when the flying circus had their heyday.

Jack was flying an antique CB Republic, near Gastonia,

North Carolina. The engine failed, resulting in Jack's death. Flying was Jack's life, his son said. Both of his sons, Jon and Charles, soloed on their sixteenth birthday.

Son Jon said "My father has more hours in the air then anyone. He was a great pilot and even a better father."

More then forty planes flew over Jack's grave when the aerial tribute was completed. One of the many standbys near the graveside remarked with a smile, "Makes you feel like Jack just flew off with one of them, all of them."

Aero Magazine, in a lengthy article stated, "Aviation has indeed lost another of its greats."

UP YONDER
In Memory of Jack Brown

In the night when the little light flickers
not and the roar of the engines are silent.
The battle of life has quit its flight and
the Pilot is cleared for service up yonder.
With a thousand beautiful things beneath his
wings. He had the best and it's time to rest.
Cleared for service up yonder.

CAPTAIN V. I. KESSLER
1973

There is no doubt in my mind that Captain V. I. Kessler is one of the most outstanding and successful pilots of West Virginia. We of Fayette Airport are very proud that his first lessons and first solo flight was at Fayette Airport. His first love is his family. His profession is flying DC-9. His hobby is small airplanes.

My thoughts go back to Kessler's solo flight. It was a wintry day with much snow on the ground. This flight was to be routine. As he was flying the airplane in the pattern, he was number two to land. This was no problem. There was plenty of time for the number one plane to clear the runway. As the first plane touched down his brakes froze not allowing the plane to move, therefore blocking the runway. A car nearby was loaded with sight-seers. Its passengers were asked to help move the airplane which was stuck in the snow. Kessler was on final with no room to land. He added the power and went around. By the time Kessler reached the touchdown point the runway was clear.

Kessler signed on with Eastern Airlines and after many years of service, became tired of the big city. He has bought a farm and moved his family where he commutes to Washington to work. He says it takes less time to fly to Washington than to drive across the city of Washington to work.

He has spent hundreds of hours preparing a fine landing strip in his backyard. Near the New River bridge not far from where he was born, another Fayette County pilot made good.

THE FATHER AND SON TEAM
1977

Bartley and Michael Blout, a father and son, purchased a Cessna Skyhawk. Bartley soloed when he was sixty years old. Michael completed his commercial course in record-breaking time and on the very day he received his license he worked several hours taking sight-seeing passengers from Fayette Airport over the Arch Moore New River Gorge Bridge.

ROY SWANIGAN

There is no one that put a student, with lack of confidence, at ease the way Roy Swanigan did. Students watching his smoothness of landings and maneuvers made it seem easy. Roy is a double amputee.

He is the first double amputee, we knew of, to hold a Federal Pilot's License.

WHEELER L. WEIKLE

At the end of World War II Wheeler L. Weikle operated a very large and successful flying school, training student pilots under the Veterans GI Bill of Rights. Wheeler was appointed Flight Examiner Pilot of the Year by the FAA. The airport and school was near Lewisburg at Boone Field.

Paul Neal, also a flight instructor, contributed greatly to the success. He later became manager of Raleigh County Airport where he served for many years.

FOR THE RECORD

On January 20, 1978, a record snowstorm hit Charleston, sixteen inches in eight hours. A total of twenty-one inches on the ground closed Kanawha Airport.

JOHN FRAZIER
1948

John M. Frazier, at the age of sixteen, arrived at Fayette Airport, starting his flying experience for business and pleasure.

It was only a few days until his father, Lee Frazier, showed up. Lee, stating, "I want to know what my son is getting himself into." I invited him for a lesson and he accepted. After the lesson, Lee said, "I do not know. I will think it over."

Lee returned two more times the same day for lessons. Leaving, stating, "I do not know whether I like this or not." This went on without interruptions for several weeks. He purchased a PA-20 Piper. The next day Lee was checked out in his plane. He asked, "How long until I can fly to Florida?" My answer was wait until morning. He did. One successful trip to Florida was the first of a long lasting adventure for Lee and his two sons. Flying to California, up the Alaskan Highway to Alaska, then to Mexico, and many trips to Canada fishing.

He, without incident, wore out three airplanes flying from Fayette Airport, before making his home Beckley where he and John, for a time, based a Cessna 210, also a 310.

I must count that father and son team high on my list of those who have helped me in the airport business.

HOBART BOOTH
Ninty-Six Combat Missions

Hobart Booth's flying career first started during his freshman year at Concord State Teacher's College at Athens, West Virginia. The flying was contracted to Princeton Airport under the Civil Pilot's Training Program, managed by Jim Tolley. Hobart's first flight instructor was Howard Grimmit. His first solo flight was in a J-3 Cub Trainer—1942.

With World War II at its peak, it was not long until Hobart was flying Douglas Dauntless Divebomber in combat, ninety-six missions in the Southwest Pacific. His plane was hit by fire four times under combat conditions.

Wherever the fighting was the hottest, Hobart was there: Bismarck, New Britain, Munda, New Ireland and Bougainville.

Hobart received a citation for knocking out a gun replacement at Tobarra Air Force Base, Spain. Hobart is also the holder of the Distinguished Flying Cross.

When the war ended Hobart came home a worker and leader in the veteran's organization.

For a short time, he added to his love for flying by using his talents for instructing at Fayette Airport.

Ambitious and restless, he, along with Rufus Hurt and Ervin Epperly, built the Oak Hill Airport where many of the present pilots of today learned to fly.

Hobart had a run of bad luck. First, on Christmas Eve, someone stole one of the airplanes and crashed it in a nearby apple orchard. Then Hobart, a few weeks later, taking off with a load of passengers in a Piper Super Cruiser, had an engine to quit. He successfully landed it in a nearby pasture field with only slight damage.

A few months later Hobart was piloting the state beer commissioner, Mr. Crow. They were taking off from the White Sulphur Airport to return to Charleston. As they were gaining altitude, the engine quit. This was in a Cessna 170. Hobart told Mr. Crow that the field which he had chosen to land in was soft and for him to brace himself. The airplane would flip over on its back. Neither Booth nor Crow was hurt. They returned to the White Sulphur Airport and chartered another airplane to Charleston.

Thirty days later, while on active reserve duty with the marines, Hobart was flying at Cherry Point Marine Corps Air Station in a Panther Jet Fighter. The airplane went into a tight left spiral at six thousand feet at 550 mph. Hobart tried to return the airplane to normal flight but something was wrong with the controls and it would not respond. Time was growing short as two airplanes circled him. He ejected at two thousand feet, his parachute opened two hundred feet from the water, with such terrific force it burst one panel from his chute. When Hobart was rescued he had multiple injuries.

Hobart retired from the Marine Corps after three and one-half years active service and remained in the active reserves until 1959. He then was retired at the rank of major.

Probably no other Fayette County man has seen the action that Hobart Booth has.

One of Hobart's big accomplishments while he was a member of the state legislature was his tireless efforts on obtaining a lake for Fayette County. The lake was Plum Orchard Lake, which should have been named Booth Lake. The lake would not have been there except through Hobart's efforts. Hobart has passed through a great adventure of dedicated service to his fellowmen—such that few men have seen.

NATIONAL FIGHT FOR FLIGHT, INC.
Not Dead, Asleep

This corporation is not organized for profit and is not organized to issue capital stock.

The charter members were: Frank K. Thomas, Roy Swanigan, James Waugh, M. W. Grimm, Kelly Skaggs, all from West Virginia and all pilots.

The purpose of the organization was to promote and encourage private aviation and to protect and promote the interests of private pilots, small airport operators and all persons interested in aviation as it pertains to private pilots; to encourage the maintenance and operation of small airports and airfields; to seek the enactment of legislation and the adoption of rules and regulations beneficial to the public in general and small airport operators and licensed private pilots in particular; to educate small airport operators, licensed private pilots and others as to all legislation and rules and regulations promulgated or proposed by federal and state agencies, and to generally protect and promote the interest of small airport operators, licensed private pilots and any and all persons, firm or corporations interested in private air flight; to acquire real and personal property incident to the operation of said corporation; to contract and be contracted with; to issue bonds or notes secured by mortgage or deed of trust on corporate assets, and to do any and all lawful manner of things necessary and incident to the conduct of said corporation.

WEST VIRGINIA AVIATION ASSOCIATION
Not Dead, Asleep

Within our time there has been several aviation organizations formed in West Virginia with nothing but good intentions. One such organization was The West Virginia Aviation Association. To name a few of their charter members who showed good intention:

Herbert Alloy, Beckley.
William C. Askew, Summersville.
Dave Baker, Burlington.
Dr. Kenneth N. Byrns, Welch.
Carlton C. Clark, Huntington.
Omar W. Grimm, Charleston.

Ray C. Epperly, Bluefield.
Lee Frazier, Oak Hill.
Dr. E. O. Gates, Welch.
Dr. J. R. Glasscock, Richwood.
J. E. Martin, Spencer.
Gerald L. Rader, Summersville.
Hulett C. Smith.
W. B. Swope.
Oscar Tate, White Sulphur.

All charter members of this organization were to promote aviation in West Virginia.

JUNE 28, 1960

The Civil Air Patrol dedicated an airport in honor of Dr. Harry Duncan. This airport is located near the village of Sanger. The Reverend Shirley Donnelly bestowed the honor.

Officers of the Civil Air Patrol were:

Captain Layton K. Pegram, commander; Second Lieutenant Seabert E. Campbell, executive officer; First Lieutenant Pat S. Pegram, adjutant; First Lieutenant James K. Pegram, training officer; Second Lieutenant Nan M. Pegram, coordinator of women; Donzetta Atha, finance officer; Claude Walsh, communications officer; Capt. Joe N. Jarrett, medical officer; Capt. Harry P. Thomas; Floyd Warrick, public information; Robert Atha, transportation and supply; Richard Noyes, Lee A. Frazier, Mrs. Elsie Frazier, Leslie Morton, Sue Koone, Gayle Koone, and Harold T. Corker.

Dr. Duncan also was presented an associate membership in the Civil Air Patrol by Commanding Officer Layton K. Pegram.

MORRIS RAYMOND "DINGER" DAUGHERTY
New Martinsville—December 22, 1894-August 13, 1964

M. R. "Dinger" Daugherty was an amazing pilot, with no legs and one arm amputated, leaving only one of his four limbs. The accident which claimed his limbs happened in 1918. He was working for the Baltimore and Ohio Railroad as a policeman. He slipped and fell and several carloads of coal ran over him. He was near death for several days in the Glendale Hospital.

"Dinger" owned four different planes during his flying career. The only known triple amputee pilot, "Dinger" was an outstanding figure in aviation of his era. In 1915, "Dinger" flew a J-I Standard, wired together with bailing wire. "Dinger's" ambition was to fly the Atlantic on the first solo flight. He was never able to promote the airplane for the flight.

He flew over thirty states. "Dinger's" talents went far beyond flying. He developed into one of the best—if not the best—dancers of the 1920s. With artificial legs strapped on, his fame was flying.

He raced cars and boats, was a musician, song-writer and author of a book.

DAUGHERTY
From the *Wheeling News*

Wheeling had the honor of being host to a number of prominent fliers during the past week. Captain Eddie V. Rickenbacker, of Detroit, assistant sales manager for the Cadillac Motor Company, in charge of LaSalle sales, spent a day in the city and was honor guest at a dinner at the McClure Hotel. He urged that Wheeling secure an airport; also, that the city mark the roofs of buildings and back all air projects.

Captain George Halderman, one of the leading aviators of the country, who, with Eddie Stinson, holds the world's aviation duration record, also spent part of two days here and took a number of Wheelingites on a flight. In topping off his stay he took Edward W. Stifel, head of the J. F. Stifel and Sons Company, his wife and Fred King of the Fidelity Investment Association for a flight over the city and then took them on to Cleveland where they enjoyed dinner and returned by motor. He was flying one of the new Stinson-Detroiter planes, of recent model.

Others to visit the city during the week were Lieutenant L. H. Scott, of Marietta, Ohio, noted stunt aviator, and M. R. "Dinger" Daugherty, "Flying assessor" of New Martinsville, who just dropped in at Moundsville to say hello to George Halderman while he was in the city.

Captain Jack R. Adams, of the Scott Field group, who is one of the directors of the West Virginia Waco Sales Company, paid a visit to Dayton, Ohio, during the past week.

Captain J. W. Hunt, of the Moundsville Aircraft Company, made a number of flights during the week.

Captain J. Orville Noll, proprietor of the Riverview Hotel, New Martinsville, West Virginia, occupied by Gloria Swanson while filming *Stage Struck* on the Ohio River, upon which Noll formerly operated excursion boats, has turned his attention to the promotion of aeronautics, and incidentally the proposed flight of Morris R. "Dinger" Daugherty, across the Atlantic.

While Wheeling, the city that sent Ruth Elder and George Halderman forth in an attempt to capture the Atlantic, was discounting the possibility of Daugherty's proposed flight, Captain Noll came to the fore with a statement that he will enter the proposed enterprise as one of a number of promoters, as a New Martinsville project.

"I expect to be in conference with Mr. Daugherty soon to learn what definite plans he has made, if any, and if there is any reason to believe the trip can be made successfully I am willing to become a member of a company to promote the project," Captain Noll said.

While Daugherty's announcement that he will attempt to fly to an unnamed city in Europe next year was discussed in aeronautic circles, the crippled birdman (he is minus both legs and his right arm) was taking special training at Langlin field,

Moundsville, under the tutorship of Captains Jack R. Adams and J. W. Hunt.

"When I leave the United States for a flight to Europe, there will be two other occupants in the plane, my favorite bloodhound 'King Brady' and a navigator, but I personally will pilot the ship," was Daugherty's comment.

"I am determined to go, and I don't care when I start or what kind of plane is supplied. I prefer an air-cooled rotary motor, possibly the Fokker model. Engineers assure me that I can have the controls set so that I can pilot the plane with one hand," he continued.

"A number of proposals have been made to me. A Pittsburgh, Pennsylvania, theatrical combine is interested and an Associated Press representative in New York City assured me that one of the large motor companies will back the flight in the spring.

"They advise me that I cannot start before May or June of 1928, but I do not want to wait that long. If the plane is supplied, I will start anytime that the promoters should decide. I would start tomorrow if the arrangements were made.

"If anybody thinks I am bluffing in this matter, let them call my hand and I will soon prove that I am in earnest. I am willing to supply a part of the capital from my savings.

"There is no question about the flight being made. That is settled in my mind. When I take off, it will be from New York to some landing field in Europe. I have not decided upon a destination. I will leave that to the promoters.

"I think it is advisable to follow the ship lanes as in the case of Ruth Elder and believe this route across the Atlantic to be the safest, regardless of it being the more lengthy of routes."

"Dinger" Daugherty, assessor of Wetzel County, of which New Martinsville is the county seat, has become notorious in West Virginia, Ohio, and Pennsylvania due to his activities as a criminologist.

With the aid of his well-trained bloodhound, "King Brady," he has tracked down many murderers and firebugs and solved a number of crimes.

CHARLES E. YEAGER

When men gather and the subject of flying is being discussed, there is one name that is never omitted—Charles E. Yeager. He was born at Myra, West Virginia, and graduated from Hamlin High School, then to enlist in the Army Air Force in September 1941.

He shot down thirteen enemy airplanes in World War II. Five of those were in one mission. He also downed one of Germany's first jet airplanes. He was shot down over France in 1944 when France was occupied by Germany. Yeager escaped captivity, fleeing into Spain.

With this vast experience after the war, duty called. To make flying safer for others, he went to Wright Air Force Base in Dayton, Ohio, and entered into experimental flying.

On October 14, 1947, Yeager made history by being the first man to fly faster than sound. The airplane was the Bell X-1. This history-making flight took place over Edwards Air Force Base in California.

In 1950, Governor Okey Patteson pinned West Virginia's distinguished service medal on Major Yeager. This presentation took place at Fort Knox, Kentucky. Governor Patteson stated that this was one of the most pleasant duties he had ever performed.

In 1973, Yeager was appointed director of Aerospace Safety for the Air Force. He has also been named to the aviation Hall of Fame.

HISTORY
Our Own Fayette County Men

This concerns the history of aviation in West Virginia.

The first airplane to ever land in West Virginia, without a doubt was Colonel Paul Peck. Colonel Peck was born at Ansted in Fayette County. He spent most of his boyhood days in Hinton.

He occasionally flew to Hinton to visit friends, and it was on such a flight that he landed at Fayetteville where the present Fayette Airport is now located.

Peck was hired to fly over the Fayette County Fair, which in those days was one of the big events of the year for the state. The fair contained rodeo races of every sort and a state convention of the Ku Klux Klan with their hoods and robes. Peck, upon this occasion, made three flights over the fair, returning to Fayette Airport, then known as Talbert Field. This established Fayette Airport as one of the oldest landing fields in the nation to still be used as an airport. This was prior to 1912. The oldest airport in the United States of America is College Park, Maryland, near Washington, D.C.

Peck gave West Virginia their first sight of an airplane. Peck was the fifty-sixth pilot in the U.S.A., and the first to carry mail in the U.S.A. The history-making flight was from Coney Island to Ohio. Peck held many records: the endurance record; the speed record of the day; and the landing accuracy record of the day. He was the first to fly over the Nation's Capital, twenty-four miles in twenty-five minutes.

Peck's favorite airplane was a Curtiss Pusher. There was no enclosure, leaving the pilot open to the exposure of the weather.

ARCHIE CLEMONS

Archie Clemons was a well-known pilot and theater owner who owned an Aeronca on floats based on the Kanawha River near Montgomery. Archie occasionally, when the river was frozen over, removed the floats and flew off the ice and visited other airports. Archie was once fined for landing a sea plane on Lunkin Airport in Cincinnati.

During a time when the Lunkin Airport was flooded over, Archie was known to be a little on the reckless side. But he was very skillful with his Aeronca. Death came suddenly when he was flying a Culver Cadet—a plane he was not familiar with. This happened at Wertz Field, Institute, West Virginia.

FOR THE RECORD
Mrs. Manila Talley

Mrs. Manila Talley, a native of Braxton County, was West Virginia's first woman licensed pilot. She was the first woman to complete the Army Air Force Officer's Reserve Course and a contestant in the 1931 National Air Races from New York to Cleveland. She was also the first person to join the Anchorage, Alaska, Civil Air Patrol, four days before the Japanese attacked Pearl Harbor.

Mrs. Talley began her flying career in 1929. In 1930 she took up racing. She also flew experimental planes for manufacturers in the 1930s. She received a graduation certificate in 1967 from the Air War College at Altus Air Force Base, Oklahoma.

Her passing came on Kitty Hawk Day, December 17, 1973. Her remains were brought home to Flatwoods, West Virginia.

JAMES K. TURBO TURLEY

James Turley was probably the first person to make his original solo flight in the border of Fayette County, West Virginia. His first solo flight was accomplished in a seaplane at Montgomery, West Virginia, off the Kanawha River. The instructor was Archie Clemons. This was the year 1940, and the airplane was an Aeronca Chief.

He soon went into service as an air cadet at Dothan, Alabama, flying a Stearm, and later to fly P-40 and P-51. Turley once had an engine failure over the Gulf of Mexico.

James assisted in the Forest Agriculture Oak Wilt Program for several reasons. He is now a teacher at Cedar Grove. He said to me, "I am anxious to retire." I told him that I never wanted to retire. He replied, "Yes, but you have never been in a hassle."

JAMES TURLEY
1964

It was late July when the day started with a usual routine. Little did James suspect that his actions would save another pilot from much difficulty. James was flying his third hour for the day on the Forest Oak Wilt Survey Program. Flying west of the Memorial Tunnel of the West Virginia Turnpike, James discovered he had a wing buddy. Another airplane followed him every turn up and down the ravines. Unable to shake the other aircraft off, James allowed the aircraft to pull up beside him. The man in the other aircraft held a map to the window and pointed to the map. James got the message. The man was lost. James headed straightway home, followed closely by the other airplane. Behind Jim the other pilot landed. He was so anxious to be on the ground that he did not even give James a chance to taxi his airplane clear off the runway. Barely missing James's airplane, both airplanes were safe on the ramp at Fayette Airport. The other pilot was thanking James and shouting praises to the good Lord. The other pilot was a preacher. Only about ten minutes of fuel remained in the preacher's airplane gas tanks, the gauges registering empty.

L. S. SCOTT

L. S. Scott, early pioneer barnstormer instructor, was one of the first to fly from Fayette Airport. He was killed May 30, 1930. The airplane was an Eagle Rock and the place was near Fairmont, West Virginia. The airplane was discharging fireworks from the air at night. The airplane spun in.

MERCER COUNTY

In Mercer County, West Virginia, flying first became a reality when James Jones and West Freemont started a flying service. This was at Bluefield, West Virginia. They named it the

Pocahontas Air Service, later to be renamed Pocahontas Air Transport. To manage the new adventure was Von Auto Krump, a German war ace. After the departure of Von Krump, Harvey Amous started a flying service. Notable flying personalities such as High Howard Wilson Dunn was on the job to assist.

Wilson Dunn related some of his personal flying history. His first solo was fifty-five years ago at Princeton, New Jersey, in a Curtiss Jinney JN4D with an OX5 water-cooled engine. His first lesson cost sixty dollars per hour. This was the year 1919. It was common practice for the student to put up a bond the value of the airplane before he was permitted to solo. Wilson Dunn is a member of the Aero Club of America, the QB, a member of the Moon Club and the OX5 Club. He has not said if he is a member of the Mile High Club. His first flying job was delivering newspapers, dropping them from the airplane to the farmers in Bland County, Virginia. Dunn remembers the first aeronautic inspectors in the eastern part of the United States.

Pop Hanskon—When Pop found an unairworthy airplane he went to work on it with his pocketknife, making sure it did not fly without being refabricated. Dunn was the owner of the first Aeronca C-2 Eve built. This was similar to the C-3 Aeronca that Ronald Thornton, from Mercer County, was killed in on Beckley-Mount Hope Airport in 1941.

During the late twenties the big day for a competition air show was at hand. Charley Lutz announced he would win the air show or tear the wings off his Kenner Fleet. Charley wired the factory asking if the airplane would take the strain of the most difficult of all maneuvers, the outside loop. The factory answered, giving the go-ahead that the airplane was structurally safe and built to take it.

The outside loop is a maneuver diving the airplane down under until it has completed a circle. This is not to be confused with the easy to do over the top loop where everything stays in place. In the outside loop the gravity pull is so great that it tries to throw you clear off the airplane. It has been known to pull the eyeballs out of socket of the pilot. Soon the hour came and it was time for Charley to attempt to rise to fame. Charley tied himself in with every conceivable belt, rope, and strap. The other pilots convinced him to use only one seat belt; in the event of a structure failure, he would stand a better chance to parachute out. Charley gained his altitude and started his down under position. We then heard the crackling sound of the wings folding beneath the undercarriage. Fear went through the hearts of those that watched. Wilson Dunn said, "So help me I raised my right hand and said, 'Good By Charley.'" But Charley was not finished. Near the ground Charley pushed himself free from the falling wreckage. His parachute opened in time to lower him gently to Mother Earth. As the first man reached Charley, he said, "I need a drink."

The story is told that only a few weeks later Charley's hair all turned white.

GLEN CLARK
July 4, 1930

At the opening of Wertz Field at Institute, West Virginia, Glen Clark was there. Clark operated a flying service at Wertz for many years, adding his talents.

Glen Clark did not quit at the close of Wertz. Glen T. Clark bought a broken-down barge in Kanawha River near Charleston for $30.00. He built it into a seaplane base and anchored it at the end of Capitol Street. You may say on the front doorstep of the city of Charleston. The city fathers and Glen had many battles, because it seems they wanted the money Clark made there in a few years. He grossed $180,000 in one twelve-month period.

The total original investment that Clark made on his first seaplane base was about seventy dollars, plus plenty of good common sense and labor of his own. His original seaplane which, in three years logged 3,700 hours, was an Aeronca, one of the first light planes to be licensed on floats. Soon he purchased three float planes. Each one not only paid for itself but earned a handsome profit. Clark's activities included flight instruction, sight-seeing trips, emergency flood relief service, photography and also sales and service of airplanes. There is no one else who gave as many persons their first airplane ride over the Great Kanawha River as Glen Clark.

In the battle of the river with the city, Clark established Clark Field, Winfield Road, Charleston, where he and Dave P. Williams operated a XGI Flying School. At an airport manager meeting, when the government was requiring the upgrading of all schools and clean rest rooms, Clark took the floor and stated he flew when the bushes and corncobs were good enough.

Clark, who had flown a total of more than twelve thousand hours, was one of the best pilots in Kanawha Valley. June 7, 1952, Clark collided with a Stinson; Clark flying a Cub. Ironically, Clark had taught Murray to fly—the pilot of the Stinson.

This closed the career of the most brilliant dedicated pilot Kanawha Valley has ever had.

I REMEMBER JAMES H. TOLLEY
1900 to 1960

There has never been among us a man with the quietness and dignity of Jim, yet a firmness which rated with the top instructors.

He moved to Pence Springs at the age of four and lived there most of his life. He operated the Hinton-Alderson Airport, beginning in the early thirties. Tolley was a commercial pilot flight examiner and his service was charter. He operated a war training program at Princeton during World War II in connection with Concord College.

Jim, at one time, barnstormed all over the state. A bit of a showman, he once hired a parachutist who turned chicken and would not jump. Jim, not willing to disappoint the public, made the jump himself. He loved people and air shows brought people. He once tied a ten dollar bill to a live goose. The goose and the ten dollars belonged to the one who caught the goose. Jim took the goose five thousand feet up then threw it out. It circled the airport several times then headed home, six miles away, landing on its own birthplace.

Jim had a great interest in the youth. Trying to stimulate greater interest in aviation among the young people, he graded off two circular tracks at the airport for model airplane flyers.

He was always proud of a visit from General Omar Bradley who was forced to land at Pence Springs in their C-47. Jim gave them transportation in his jeepster. Shortly after landing, the C-47 took off for its base in Washington, D.C.

JIM TOLLEY
1942

Jim was finishing his second year of operating the Civil Pilot's Training Program at Princeton, West Virginia.

As we have stated, Jim was the most mild-mannered person ever in the airport business. There comes a time in the life of every man when he has taken all he can stand. Jim actually kicked the rump of a federal inspector, causing no damage to Jim's foot. The school was immediately closed. An investigator came and decided if there was any publicity this might be a common practice. Jim, with his super, well-mannered disposition, convinced the inspectors he had done his good deed for the day. His school reopened.

CHARLES LILLY

In the late twenties and early thirties, if an airplane was flying, Charles Lilly was there. Lilly, a lover of adventure, was there for the love of flying and to help others, not for money. All of his spare time was spent around the pioneer pilots.

Lilly said that in those days, the late twenties, a man was a daredevil for piloting an airplane. In those days, Lilly's favorite airport was Tolley Field, near the present location of the Pinecrest Sanitarium. John Tolley leased the field to the old barnstormers, as they were known in those days. Those pilots came in every aircraft known of in that day, Tri-Motor Fords and Stinson.

Lilly said that some of those airplanes were hardly more than flying chicken coops strung together with bailing wire. If it broke down, we were not too particular how we repaired it. The regulations governing the repair of airplanes were not as strict then as they are today.

Charles Lilly, once manager of the Beckley-Mount Hope Airport, with his labor of love, contributed much help and inspiration to pilots and operators of airports.

GILL ROB WILSON

Of all West Virginians that have made their name in aviation history, Gill Rob Wilson of Parkersburg brought joy to the hearts of all that knew him and read his editorials.

His constant visits to Parkersburg, flying his own light plane, to visit his mother and father who lived there, were always welcomed.

In 1930 he was appointed as director of aviation for the state of New Jersey. On weekends he joined his closest friend, the famous Billy Mitchell of Long Island, to fly with the Reserves.

In 1940 he was the New York Herald Tribune aviation columnist. During World War II, in European and Pacific theaters, he flew with every air force on every front line and cabled the stories back to his paper. He called himself the World Wide Prowler. He had traveled in every country in the world and after thiry-five years of travel, he said that he was tired and wanted to land somewhere.

He accepted the position of editor of the *Flying Magazine*, the only independent voice of aviation. His years of aviation touched on nearly every branch of the profession. He was chairman of the first aeronautical division of the American Legion and consultant to the federal government on Lighter Than Air Development. He was co-founder of the Aircraft Owners and Pilots Association. He was a member of the committee which developed the Civilian Pilot Training Program and in 1938, he created the plan for the Civil Air Patrol.

He was also vice-president of the Air Force Historical Foundation and a close friend of General Carl Andrew Spaatz and Jimmy Doolittle. He happened to be at Lakehurst, New Jersey, in 1936 when the Hindenburg burned. He was shot down from ten thousand feet in World War I.

His biggest thrill was flying home and getting sight of America again.

We of West Virginia are proud of one of such fame. Wood County Airport is now so named Gill Rob Wilson Field, as it should be.

ROBERT THOMPSON
Adventurer

Robert Thompson, after graduation from Elkview High School, started flying at Clark Field in Winfield, instructed by Dave P. Williams. Thompson worked at the airport to pay for his flying.

After serving in the air force, he spent twenty-two months in Korea, then came back to Charleston to serve as a pilot to Governor W. W. Barron. After one summer of flying on the State Oak Wilt Survey, Thompson went to Virginia as a crop duster. He then returned to Charleston as flight instructor at Kanawha Airport. Piedmont Airlines offered him a job after he got his commercial license but he turned it down saying he didn't want to know where he would be going every day.

Bob then found excitement as a charter pilot in Florida. From here much mystery surrounds his work and little is known, except that he was working with the Anti-Castro Forces.

Thompson told friends that he had made several flying trips over Cuba, dropping leaflets, guns, and dynamite to Anti-Castro Forces. He was first reported missing while en route from Fort Lauderdale, Florida, to Nassau.

A Cuban refugee told the story that Thompson had crashed. He said that he had been shot down while making a low pass over the city and his plane hit a sugar mill.

Other stories reveal that he was making a landing to let off a passenger in Cuba, being shot and wounded. Does he live or not? We do not know.

I personally knew him and neither words nor pen can describe his personality and enthusiasm.

STAFF SERGEANT JAMES K. HALL

It is with pride that we record the deeds of one we were so close to.

Staff Sergeant James K. Hall distinguished himself by outstanding achievement as a flight engineer technician, helicopter, on October 1.

On that date, Sergeant Hall modified and tested a Stokes litter to be used for retrieving a critically ill crew member of the U.S.S. *General Patrick*. Upon arriving at the area of the *General Patrick*, bad weather was encountered. Rain, ocean swells up to twelve feet, and white caps caused the large ship to pitch and roll excessively. Regardless of the hazards involved the crew went ahead with its mission. During the rescue operation, Sergeant Hall provided the pilot with precise position instructions and expertly operated the hoist to keep the litter clear of the ship's superstructure. With Sergeant Hall as flight engineer and hoist operator, a successful pickup was made. The critically ill crew member of the *General Patrick* was returned to Tripler Hospital and has completely recovered.

By his demonstration of outstanding professional skill and crew coordination, Sergeant Hall has reflected credit upon himself and the United States Air Force. On the basis of his lifesaving mission the air force commendation medal was awarded to crew member Staff Sergeant Hall.

He was a 1955 graduate of Collins High School. He was also a member and leader of the Air Explorers Division of the Boy Scouts of America in which Frank K. Thomas was the adviser at the Fayette Airport.

HOWARD CLIFTON "TICK" LILLY
1948

Howard Clifton "Tick" Lilly, a true pioneer in jet aviation, departed from us at the early age of thirty. A letter to his parents from the head of the United States of America Test Unit paid this tribute:

> His contribution and sacrifice for the Nation was as great or greater than that in the Armed Services of this Country during the war and he rightfully deserves a place with Colin Kelly, Booth O'Hare and Richard Bong.

The work he was doing in high-speed fighters was pioneering of the same type as the first flight of the Wright brothers. He will not be forgotten. "Tick" was awarded the air medal for heroism in aviation.

"Tick" was to realize his ambition—to travel faster than man had ever done. It was only a few months before his death that it was announced that he had flown through the supersonic barrier which required a speed of 762 miles per hour. "Tick" and another West Virginian, Capt. Charles Yeager of Hamlin, were among the first known pilots in the world to have flown faster than sound.

In May 1948, "Tick" took off in the Skystreak. At two hundred feet the engine exploded and the plane rolled over and crashed.

His flying career began at the Beckley-Mount Hope Airport where, in the spring of 1941, he took up flying with Karl Williams as instructor.

He, in such short years, accomplished more than most pilots do in a lifetime.

LADY FLYER

Mrs. Ruth Tolley Gwinn, daughter of the aviation pioneer James H. Tolley, and the wife of John W. Gwinn, the manager of the Greenbrier Valley Airport, was the most outstanding lady pilot of West Virginia.

Mrs. Gwinn offered her services to a Florida couple who landed a Cessna 150 on a hillside farm during a bad rainstorm. Mrs. Gwinn was to fly the Cessna from the hillside. Probably due to a wind switch, she found it impossible to clear the fence, therefore putting the airplane on its back. Mrs. Gwinn, unshaken, proceeded to the Greenbrier Valley Fair where she was to meet her husband John. When he met her he asked, "How did you make out with the little Cessna?" "Oh, I tore it to pieces," she said, "I'll tell you all about it later."

Mrs. Gwinn has put in thousands of hours of instruction throughout southern West Virginia.

COLLAPSE OF HANGAR'S ROOF UNDER SNOW CRUSHES PLANE

It was "ceiling zero" for an airplane here this week.

The roof of a hangar at Boone Field fell under the weight of snow and crushed a light plane belonging to Pence Airways of Pence Springs.

The craft, a Cessna Skyhawk 172, had been in the hangar since Col. and Mrs. John Gwinn and sons returned from Florida last week.

Gwinn placed the airplane in the hangar at Boone Field because the Pence Springs airport was snowed in when he arrived.

WEST VIRGINIA'S MOST EXPERIENCED PILOT

by Adrin Gwin, *Daily Mail*

He's the pilot.

He owns the charter planes, and he flies them to take people to and from the Greenbrier Hotel in White Sulphur Springs.

Oscar Tate is Greenbrier Airlines. He and about a dozen other people. But Oscar is the founder, owner, manager, chief pilot, and skycap-porter for the airline.

Sometimes he's the ground-operations crew, and sometimes he's the traffic controller at the Greenbrier Airport, and other times he's ticket agent at the front counter.

And all the time he's all of those things, for Charles Oscar Tate, Jr., *is* the Greenbrier Airlines.

He's been flying for forty-three years—most of it over West Virginia's rugged hills and steep valleys—and he isn't thinking about quitting.

Flying is his life—and we misunderstand if we think he flies just now and then for fun, like most sixtyish, quiet-mannered but energetic businessmen might. He flies because that's all he ever wanted to do.

Oh, back in college at Marshall down in Huntington he took journalism under Page Pitt, the legendary blind instructor who retired several years ago. But he wanted to fly, so he did.

"I traded a shiny new Model A Ford and $250 for flying lessons, and soloed in 1932 at Huntington," he said. (He was born there, and Howard Mays, one of the early West Virginia flying instructors, taught him.)

He wanted to get into the air force, but so did about seven thousand other young men. Languishing on a waiting list wouldn't put him in the air. Waiting wasn't his bag. He soloed in 1932, got his commercial license for small aircraft and ran a flying school in Huntington and Barboursville from 1936 to 1939.

In 1939 he went to White Sulphur Springs to start the Greenbrier Airlines. "There was one hangar, and one old green Stinson airplane here at the time," he said. "That was it."

Today, Greenbrier Airlines has eleven planes, all of them sleek, fancy, fast, and efficient—especially with Capt. Oscar Tate driving them. Some of them will hold seven passengers, some of them are three-passenger planes. But Tate and his crews can take them from the Greenbrier to wherever you want to go, at your command.

With the old "seat-of-the-pants" skill acquired in the early thirties before sophisticated instruments, Tate can swing a little Cessna down onto any river-valley landing strip. And he can take his place in the landing patterns of the biggest airports with equal aplomb, running through the complicated procedures of specialized jet-age flying with the computer-like control and calculated efficiency of an Air Transport Command pilot taking a giant liner across the oceans of the world.

That's what he did in World War II—flew Trans World Airlines air transports from the United States. Across the North Atlantic to Goose Bay, Labrador, Iceland, Greenland, England, and Scotland. Across the South Atlantic to the Azores, North Africa, and Italy.

He's been back at the Greenbrier since 1946, building up his beloved Greenbrier Airlines.

And flying. Always flying. Call the Greenbrier Airport number and someone will likely tell you that he's out right now.

"He's just landing at Roanoke on a charter flight," they'll tell you. "Be back here in about thirty minutes if you care to call back."

If the weather's good and the traffic is heavy, Oscar might fly as many as five flights a day from White Sulphur Springs—to Roanoke, Charleston, Pittsburgh, Morgantown, and perhaps even one to Washington—and be back home before dark.

Meanwhile the five other pilots of Greenbrier Airlines might be doing as many flights each at the same time. "Our season is the Greenbrier season," Oscar said. "We go where they want to go, and when they want to go."

GREENBRIER PILOT WINGS BACK INTO HISTORY
White Sulphur Springs' 'Red Baron'
by David McCorkle, *Daily Mail* staff

Greenbrier Airport Manager Oscar Tate is winging himself back into history with his recently purchased 1928 American Eagle biplane.

Dressed in knickers, helmet, goggles, and a leather jacket, Tate is becoming known as White Sulphur Springs' "Red Baron."

Tate recently found and purchased the 1928 airplane relic under rather unusual circumstances. In February he was attending the funeral in Huntington of Howard Mayes, an area aviation pioneer who taught Tate to fly in the early 1930s. While at Huntington Chesapeake Airport, Tate was shown the 1928 American Eagle, one of the planes which was buzzing fields when Tate learned to fly.

The plane belonged to the widow of an Ohio pilot, Leonard A. Fry, and had been brought from Portsmouth, Ohio, to be hangared at Chesapeake.

Tate was intrigued with the intrepid two-seater and purchased it shortly thereafter. It had been restored in August 1967 and was in remarkable condition. Tate found the original logbooks in the plane with entries as far back as 1929.

One logbook contained a letter, several decades old, to the owner from Victor Roos, one-time president of American Eagle-Lincoln Aircraft Corporation and an aviation buff and financier.

In addition, the books revealed that the craft had been in storage for thirty-two years—from November 1935 to August 1967. The books gave no reason for storage, however, and Tate's efforts to find why it was not used have met with negative results.

Tate, who doesn't believe in putting his candle under a bushel, demonstrates his biplane each Sunday at 2:00 and 6:00 p.m. at the airport. He takes the craft, which cruises at seventy-five miles per hour, through given figure eight maneuvers, stalls and dives for interested onlookers.

He is also very much interested in the background of older planes and explains the circumstances surrounding the odd-looking extended nose of his craft:

"The close of World War I left as surplus thousands of the airplane engines which had never been used and because they could be bought cheaply, airplane models were designed around the engines.

"Time, however, brought new developments and engine companies perfected an engine with greater horsepower and less weight.

"Airplane designers wanted to use the new engine, but found it economically impractical to redesign their models. As a result, they merely extended the plane's nose to accommodate the engine and its reduced weight."

News of the little plane and the "Red Baron" is spreading and jet pilots occasionally drop in to see the vintage craft, Tate said.

He believes the plane will be a drawing card for the airport as well as for White Sulphur Springs. And the community has welcomed the craft and pilot as a durable duo they can boast of.

GODSEY
1930

In the early thirties, and near the sight of the present Fayette Airport where the Emmanuel Baptist Church now stands, some local flying enthusiasts cleared the brush. They then hired a plane and pilot and had leaflets printed. Pilot Ernest Godsey was to distribute the leaflets over Scarbro. He had engine failure and upon landing the plane, hit a tree stump but with apparently little damage. After repair to the engine he was off.

The next day being Sunday and the day of the big air show, the people insisted on aerobatic Godsey. At first he was reluctant but then decided to please the public. Starting his first loop, the airplane shed a wind and spun hard to the ground, demolishing the airplane and leaving Godsey with an injured back, from which he recovered.

Later that month, Mr. Godsey persuaded Bill Layton and Harry Berry to finance an airplane. This airplane was located at Narrows, Virginia. Upon takeoff, the engine failed and the airplane landed in a shallow place in New River, flipping on its back. It pinned Godsey beneath the airplane, but two young men were able to hold Godsey's head above the water until help came. Godsey had no injuries, but the airplane was a loss.

Godsey later spent many long successful years with the Civil Aeronautics Administration (CAA) and is now working for the FAA. Mr. Berry, the financier of the airplane, until his death, had the same feeling toward airplanes as McGovern did Nixon.

THAT'S MY CAR
1942

B. L. Slater, an old-time flyer, spent his weekends instructing and flying passengers at the Beckley-Mount Hope Airport. This event happened as B. L. Slater was hauling passengers in a 1929 Curtiss Robin Airplane. This airplane was the same model as the one Wrong Way Corrigan flew to Ireland. It was up on takeoff. The instant Slater pulled the airplane into the air the entire left landing gear collapsed, giving way and hanging loosely beneath the airplane.

This could mean that upon landing the aircraft could go on its nose or back with the gas then spraying over the hot engine. Need we say more?

H. L. Sessler was at hand. He jumped in the first car he came to, which happened to be my car, and attempted to line up with the landing aircraft by maneuvering my car under the wing of the landing airplane to substitute for the broken gear. This was a tense moment. Slater was more skillful with the old airplane than Sessler was with the car. Slater landed with one wing high on one wheel, using a little power holding the damaged gear off the ground until nearly all the speed was gone. There was very little damage other than the give way on the lift-off. The airplane was a tired old bird but with very little work it was ready to fly again.

I REMEMBER H. L. SESSLER

As I remember, H. L. Sessler, he was hard working, sincere, dedicated to his profession: a very independent, rugged individual. He was always happy, and always ready to tell or hear a joke, with a lasting sense of pride for his work. Sessler gained the respect of all that knew him.

He enjoyed telling the mistakes he himself made, and it is with respect that we record some of these.

Sessler, who taught hundreds to fly, learned flying from Bill Duff of Bluefield in the late twenties. He had to put up bond for the value of the aircraft before he was permitted to fly it alone.

Sessler was a mechanic in the Army Air Force in France during World War I, bringing home his skills. Throughout his life he benefitted southern West Virginia aviation enthusiasts. He built and rebuilt hundreds of airplanes.

One project brought him much distress. After building a Parasol Heath Airplane, he was to start an airport on the Fayette-Raleigh County line, to be called the Fayroal. Sessler advertised and told the public that they would see this little home-built plane fly for its first time at the grand opening of the field. The evening before, though, he could not resist the temptation. He flew the plane, cracking it up on landing. It took months of hard work rebuilding it. Sessler flew the frail craft to Bluefield, leaving as darkness began. The craft spun out. Feeling he would not be found, he started his journey walking down a long dirt road with minor injuries, except a large part of his nose being cut off. Feeling that he would faint, he washed his face in mud holes along the way.

Returning to Mount Hope to run a garage and rebuild old airplanes, he supported from the rafters by a chain a complete aircraft engine. Passing it daily, he dreamed of the day he would be up there behind the engine. He grasped the propeller with both hands as though he were starting an airplane by hand. The magnetos were not grounded and there was gas in the carburetor causing the engine to start. The propeller cut the chain and the machines danced throughout the shop, damaging three cars and chasing all employees from the shop before it ran out of gas trying to climb a wall.

FOGGY—1941

Sessler's moonlight sight-seeing ride with his Waco departed Beckley-Mount Hope Airport with three passengers. The men were anything but sober. They even had to be helped into the airplane. But this did not disturb Sessler because their money was green.

He was off, to fly over Oak Hill's night football game in progress, returning to Beckley-Mount Hope Airport. They found the airport fogged over completely and they had passed the port of no return. There were no lighted fields in the area and if they tried for one, the condition may be worse. Win or loose, Sessler must try.

His first try he let down where he thought the field was. Brushing trees, he pulled up into the starry skies above the fog. On the second try, he could see an outline of the airport but he was too far down the runway to land. Up into the starry sky again.

It was time to tell his passengers that the situation was bad. He had taken up three drunks, but to his amazement, it was easy to see that he now had three of the most sober men in the country.

The third try was a perfect landing.

WHITE SAND

One of Sessler's favorite jokes. He asked all that entered his office, "Where can I find some white sand? I must have pure white sand." His visitor would then ask what for? Sessler's reply, "For my white cat, of course."

This went on until someone sent him fifty pounds of sand C.O.D. air freight from Daytona Beach, Florida. Thinking that the sand was airplane parts, he paid the C.O.D.

That was the end of the cat, because it cost too much to keep him.

BAMBOO BOMBER
1941

The government was taking all good training airplanes from their owners for their primary cadet flying program. H. L. Sessler's Flying School turned over three, and the last was to go soon. To save the last J-3 Cub, he removed all good parts and replaced them with worn-out, bursted ones, making the Cub unacceptable.

Waiting until the shortage of airplanes ended, he restored the little bird and put it back into the air, this Cub being the only one of its make flying in the area.

Sessler, having sacrificed three aircrafts, convinced our senator that he should have priority over others for surplus. In 1944 they gave him the pick of the Twin Cessna UC-78 Bobcat, otherwise known as the Bamboo Bomber. Some called this craft the Match Box.

Sessler was ready for charter. He hired a young pilot from New York. When Sessler met the train, the pilot wished to see Beckley-Mount Hope Airport before going to town and hunting a place to live. He asked if he could fly the Bobcat. Sessler agreed. Upon landing, he forgot to lower the landing gear, damaging both propellers. Sessler drove his car to the airplane and picked up the pilot, taking him back to Prince to the train station. Sessler did not wait to see him off.

Months passed, waiting for parts to restore the aircraft. On the big day, Sessler taxied out for a test flight and during the preflight run-up, the gear was not locked causing the plane to collapse again.

RETRIEVE

After the passing of Sessler, his family, living in Florida, asked if I would purchase his equipment and remaining belongings stored at Beckley-Mount Hope Airport. I was to return any personal belongings to his family. I found that several people had been taking what they wanted—a little at a time.

For the removal, it seemed best we surprise the attendant, thus preventing any further looting.

When the day came, we were waiting on the attendant to unlock. We had three moving vans. We were met with much protest. I explained I had purchased all aircraft parts, supplies and shop equipment but would leave enough of the furniture that business could be continued.

They surprised us with a distress warrant, meaning we would have to leave it alone or give bond until a hearing. I went to town and gave bond in order to remove everything. I told them to remove their papers from the files or they were going to Fayetteville since I would be moving the office furniture first, including all chairs. Twelve truck loads went to Fayetteville. We took everything that was not nailed down except the secretary and telephone.

Beckley-Mount Hope Airport was due a visit from the federal inspectors the next day. The federals, thinking we had pulled a dirty trick, thought they would find my place cluttered. Upon arrival to Fayetteville, they found our house in order. We had stored all truck loads in a warehouse.

Weeks later, a man came by claiming a large wooden cabinet. I contacted the Sesslers and we believed it to have been part of Sessler's estate. This man demanded its possession. I gave him a chance to run or go back in it. When we discovered he owed the mechanic seventy bucks, we ran, caught him, and he cheerfully paid?!

FEBRUARY 9, 1969

The Aviation Award was given to Dr. Enrique Aguilar of Charlton Heights. This award to Pilot of the Year was recognized at a dinner meeting at Fayette Airport with two hundred persons attending. It was a snowy afternoon and there were pilots from throughout the state.

Dr. Aguilar, to our knowledge, is the first to receive such a certificate of appreciation for his support, work, and enthusiasm in the field of aviation.

He is not only an outstanding pilot, but also aviation medical examiner. Judge Charles L. Garvin, Jr., presented the award.

LEWISBURG—1949

On July 6, 1949, Wheeler L. Weikle, the man that contributed much to the growth and interest of flying in Greenbrier County, had his unlucky day.

With Weikle as flight instructor, R. H. Bowman, a student from Rainelle, received only minor injuries when the Luscombe Airplane failed to maintain sufficient power after takeoff. The plane gained approximately twenty feet, then immediately started losing power. There was not sufficient space to land on his own runway, so Weikle attempted to land in an adjacent field.

The plane skidded to a stop against a fence and was completely demolished.

For The Record

FOR THE RECORD
September 3, 1921

The first aircraft accident I have knowledge of in West Virginia occurred September 3, 1921, at Drennen, West Virginia.

There was bad labor trouble throughout the coal fields as the union was being organized. Some thought there would be a revolution at Logan, West Virginia, as there was a gathering for an armed march of miners. President Warren G. Harding ordered soldiers into Logan, also sending twelve bombers to fly over the area with the rumors the airplanes were carrying bombs. There was never a showdown between the army and labor. The aircrafts all became lost. None reached its home base without first becoming lost and having to force land. The worst of the crashes was at Drennen where four were killed and one injured. It was three days before the wreckage of the bomber was found.

WEATHER

Looking at the oldest weather forecast on record, we find recorded in the Holy Bible: Job 37:22, "Fair weather cometh out of the north."

Matthew 16:2, 3, records, "When it is evening you say it will be fair weather if the sky is red. And in the morning it will be foul weather today for the sky is red."

My comment is to watch these signs. If you have had several days of bad weather and the wind switches to the north, watch the barometer rise and the sky clear. The prediction from Matthew never fails.

TO SEE OR NOT TO SEE

John Eye, of the Oak Wilt Division of the Department of Agriculture, was my observer on a survey flight to search for blighted trees. As he pointed down, I was to make a low flight over the area in order for him to take a more careful look at the tree for positive identification. After fifty minutes of low flying I finally told him, "There is no blighted trees there." He said, "Who cares? You are not looking in the right place. Look on the old strip mine. There are three girls taking sun baths." Mr. Eye said, "I have the way mapped in. I'll take sick leave for the rest of the day." And so he did.

C. E. MARTIN
1930

C. E. Martin, in an early model Kenuck, while flying near Beckley, developed engine trouble, putting the airplane into a tall maple tree. Using tall ladders, the Beckley Fire Department rescued him. Both tree and airplane received major damage.

In the early days of his flying, Martin frequently used a well-placed but loosely piled haystack to stop his airplane while landing at Fayette Airport.

JULY 2, 1942

A converted American Airline Flagship crashed in flames into a garden at Premier, near Welch, West Virginia. All twenty-one soldiers on board were killed. The airplane had been flying above the clouds. The pilot radioed an accompanying airplane that he would descend beneath the overcast.

APRIL 8, 1951

Under instruments, bad weather, and low ceiling conditions, twenty-one West Virginia Air National Guardsmen were killed as a result of the crash on a hillside near Kanawha Airport. Notice the similarity to the military crash near Welch in 1942. Each crash claimed twenty-one lives.

AUGUST 10, 1968

At 8:56 a.m., a Piedmont airline crashed at Kanawha Airport. Thirty-five were killed and there were two survivors.

NOVEMBER 14, 1970

At Kenova, West Virginia, near Huntington, under instrument, bad weather, and low ceiling, a DC-9 Southern Airline, carrying Marshall University players, coaches and fans, crashed. Seventy-five lives in all were lost. There were no survivors.

WITHOUT WING TIPS DC-3
May 2, 1943

The word came first to Beckley-Mount Hope Airport that a DC-3 Transport had structure failure and crashed in the Flat Top area south of Beckley. It was first believed two aircrafts had collided. Six feet of the plane wing tips broke off with the aileron found immediately.

The airplane's home base was Roanoke. The pilot declared an emergency. The pilot had been practicing unusual altitude under instrument conditions which caused too much stress on the airplane—probably excessive speed in the recovery from stalls.

With both wing tips and other wing controls gone, the pilot found the airplane to be controllable with rudder and alternating the power of the engine.

To return to Woodrum Field in Roanoke area would put the path of the airplane over the thickly populated areas. In the event of complete loss of control, the casualties would be high.

The pilot's decision was good. He headed for Pulaski Airport, which was a sod field. This would help prevent fire in the event of a ruptured gas tank. Reaching Pulaski Airport, surrounded by farmland, he belly-landed it. The pilot and the co-pilot were uninjured. The DC-3 was repaired and flown from the airport weeks later.

BAIL OUT

On February 8, 1942, at 12:15 p.m., Thomas W. Gillispie of Oak Hill, West Virginia, a student of VPI Airport, Blacksburg, Virginia, was flying with an instructor in a Piper J-5 Cub when the controls became locked. The instructor, C. W. Watrous, ordered Gillispie to bail out at one thousand feet. Gillispie followed orders and was only bruised from the experience.

The instructor tried to bring the plane down, without success, and from an altitude of 250 feet he jumped. The altitude then was too low for his chute to open and he was killed.

The plane landed on its wheels, not badly damaged. The coroner expressed belief that the instructor would not have been hurt if he had remained with the plane.

Gillispie finished his training and flew in service during World War II.

DOPE

I attended a lecture on the detection and identification of narcotics. This lecture was for the purpose of safety, helping us to keep a careful eye on our young flyers.

The speaker stated that the symptoms were dilated pupils of the eyes, thickness of the tongue, and slowness of speech. The charcoal color and sweet odor of newly mown hay is like the color and smell of marihuana. There is also a white frost-look ring around the mouth. I had constantly noticed this white ring around the mouth of one of my students, but his actions were normal. It was time to keep a close look for other signs with this boy.

I then noticed the same was appearing around my mouth—it was toothpaste.

NIGHT RIDERS
1946

It was a dark night and there were late customers waiting for rides. I asked Willard Lowery to hold a lantern in the center of the airport in order for me to see where to land. It was so dark he could hear but not see the airplane. As the sound grew closer he took off running for the woods. He did not want that propeller cutting his hair. I chased the lantern for a short distance, then circled again. Being familiar with the field I landed with the light of the kerosine lantern.

TWELVE-YEAR-OLD SAVES TWO IN PLANE CRASH
1948

September 25, 1948—A Piper J-5 Cruiser flown by Donald Yost, Speedway, Fairmont, West Virginia, was injured, Homer Riley, Cleveland, Ohio, was injured and A. V. Clevenger was drowned when the ship they were flying in crash-landed in the Monongahela River near Hoult Locks.

Tony Depapas, twelve years old, is a hero. Tony's quick thinking saved the lives of Yost and Riley. Tony saw the plane crash into the river and immediately pushed out in a rowboat in an effort to save the survivors. He reached one man but could not haul him into the boat, so he just rowed to shore with the man clinging to the side. Older men took over from there, so we are told, and rescued the second man.

THIS MAY BE A FIRST
1948

Three solos in one day is not unusual, but this time there were three from the same family, each not knowing the other had soloed.

First, G. Lee Massey soloed, flying his airplane alone for the first time while his two sons were working on their J-3-C Cub in the shop. He asked that I not tell Larry or Jerry that he had made his first flight alone. Lee left the airport for the day.

Next was Larry's turn, leaving his brother working in the shop. When Larry had finished his solo, I asked him not to tell his brother he had accomplished his first solo.

Then his brother Jerry's turn had come.

Each of the Masseys, not knowing the other had soloed, flew alone for their first big thrill that can only be explained by those that accomplish it.

TELL US WHY
1948

On August 6, Dr. R. J. Dahney and Jack Adams, both of Taylorsville, North Carolina, were killed when their plane, a Navion, crashed in a dense wooded area not over three hundred feet from the end of the runway at Kanawha Airport. The pilot radioed to the tower about 6:55 that he had only about eighteen minutes of gas. The plane went out of sight and was not seen again until it was found Tuesday, August 10, by a person picking blackberries.

Searchers had scoured the area for more than fifty miles in all directions from the airport. More than one thousand flights had passed over the scene of the crash.

It was said that the doctor refused to pay a landing fee at a large airport. Therefore, they refused to sell him gas. He believed his tax money built the airport so why should he pay a landing fee?

NOVEMBER 20, 1948

Anthony Pizzaferato, Norton, West Virginia, was fined twenty-five dollars and costs when he entered a plea of guilty before Justice of the Peace G. W. L. Doyle of Elkins, for low and reckless flying. Pizzaferato, the defendant, was caught flying low in a reckless manner over Wimer Stadium, at Elkins, during the Davis and Elkins and Wesleyan College football game. This occurred on November 20, 1948.

On November 20, 1948, Charles Kesner of Nutter Fort, West Virginia, was fined twenty-five dollars and costs when he entered a plea of guilty before Justice of the Peace Charles D. Smith of Clark District, Harrison County, for reckless flying.

On November 4, 1948, Homer William Aliff, Princeton, and Randolph Blankenship were given minor aid at the Princeton Memorial Hospital for injuries received when "shortly after take-off at approximately 2,000 feet altitude, the pilot observed the engine cowling coming loose. He throttled back and turned the nose to level flight. After completing the 180° turn, he applied throttle when the engine failed to take, resulting in a forced landing causing damage to aircraft by going through fence."

FEBRUARY 10, 1951

Clyde Tinley, a pilot from Akron, Ohio, and two passengers, were downed in Jackson County on the Organ Farm when they became "lost" while flying between Parkersburg and Charleston. They tore down some fences; but no one was injured. They flew the plane out the next day.

FEBRUARY 11, 1951

Aaron Cutlip, pilot, landed his plane in a field adjoining the Morgantown Municipal Airport. His motor quit while on base leg and due to high winds he had to make the emergency landing. There were no injuries, and only slight damage to the plane.

REGISTERED MAIL
1958

Registered mail from the bureaucrats usually means bad news. This particular letter stated that Piper N5561H, registered owner Frank Thomas, was believed to have been flown to Woodrum Field, Roanoke, Virginia, on or after the expiration of the annual inspection.

The letter stated, "Please give from your records the last date flown, the pilot's name and why you had not secured a Ferry Permit." My reply was, "I have searched the records and have found no evidence of this aircraft being flown after expiration date of annual."

P.S. Please tell me. Are ferry permits issued by fairies? If we need a permit, where do we catch a fairy?

GIFTS

Rev. E. E. Hale brought me a box of instruments from a small airplane as a gift. I asked the history of them. The Reverend Mr. Hale said he was once the owner of an Erocoop. As he was traveling through Indiana, the weather became so bad it was necessary to land in a pasture field. This would have offered no problems except a bull ran out in front of his airplane as he was touching down. It was hard to say which was in the worse shape—the airplane or the bull. The airplane junked on the spot and the bull processed for hamburger.

RAINELLE
1950

On May 5, Lacey Manspile was fined one hundred dollars for his determination to fly without the permission of his flight instructor. Manspile pleaded guilty before C. L. Craft, justice of the peace, for operating an airplane without a license.

State Trooper C. R. Dawson told the story. Manspile was taking lessons from a flying school near Rupert. The instructor stopped Manspile, telling him that he could never learn to fly. His lessons were halted.

Manspile asked if he could taxi the craft around on the ground a bit. Taxiing too fast, the plane hit a bump and up it went. Manspile was off for his first solo. After one hour of enjoyable flying, Manspile landed, taxied to a stop and walked away smiling.

FROM THE SCRAPBOOK
July 4, 1960

This incident takes place in Hopewell, Virginia. C. C. Coffman, making his first parachute jump, dangled helplessly thousands of feet above the ground for two hours when his parachute snagged on an airplane.

Coffman finally was freed when the pilot, Mrs. K. R. Bradley smashed the glass of an instrument dial with her shoe using the broken glass to cut the shroud cords. Coffman pulled the emergency parachute and floated down, with only slight injuries.

WATER
1943

I was leisurely resting, reading in the office of the Beckley-Mount Hope Airport. A quite slow, soft-spoken man, his action and breath giving evidence of alcohol, approached me and said, "Mr. Thomas, when you are through reading, I want to talk with you." To hasten his departure, I replied, "What do you want?" He replied, "Water." I told him that would be something new for him. He said, "You don't understand. I want a bucket full. Where is your bucket? I ain't got none and that's why I come to you." I asked, "Why do you want a bucket of water?" He said, "Turn around and you will see. My x ! x ! car is on fire."

All the airport help was immediately summoned. It was a cigarette fire consuming the back seat of his car. It had not yet reached his gas tank. His car was filled so full of water that it came out the windows.

FROM THE RECORDS
October 3, 1949

Russell E. Ferrence, private pilot of Alvon, West Virginia, was uninjured when the plane he was flying, owned by C. O. Tate, Jr., of White Sulphur Springs, collided with a wire fence, gate, and locust post which separated large fields he was using for landings and takeoffs from another field.

Ferrence had taken off from Charlottesville, Virginia, on a return flight from Charlottesville to White Sulphur Springs. The pilot stated the reason, "The engine was heating up and losing power. I hit a fence while attempting to takeoff because horses were running toward the airplane." The owner of the airplane believes the cause of the accident was due to landing for personal reasons near his home in an unauthorized field and not due to malfunctioning of the airplane. The landing was one-half mile from his home at Alvis, eleven miles northeast of White Sulphur Springs.

FROM THE RECORD
January 1, 1951

Fred Clark, a pilot from Dunbar, became confused, after leaving Point Pleasant to return to South Charleston, in finding the Ohio River to his left. Clark thought he had crossed the Kanawha River at this point without knowing it. He kept bearing left thinking he would again cross the Kanawha, but instead, finding himself lost. Darkness was falling and the plane was running out of gas. Clark decided to land his Taylorcraft equipped with pantoons on the road. The plane was scrap, but there were no injuries. Apparently seaplanes do not land easily on roads.

June 12, 1949

On June 12, 1949, J. J. Farrar piloted a civil aircraft carrying a passenger. He took off from Hinton-Alderson Airport at Pence Springs—directly behind a twin-engine Cessna. He encountered the prop wash of the Cessna, which caused Farrar to lose control of his aircraft and crash. Farrar, operator of the aircraft, was considered careless by the federals. Farrar was given a warning. No other action will be taken at this time.

August 22, 1948

Allen Pennybacker, age twenty, and Billy Marlow, age seventeen, both of Parkersburg, struck a high tension wire about three hundred feet above the ground, scraped the top of a small sycamore tree, and crashed head on into the ground, bursting into flames. Their broken and charred bodies were removed from the wreckage approximately an hour and a half after the wreck. Witnesses related that the plane had been flying at a low elevation over the scene of the crash for several minutes. The scene of the crash was a picnic area. The pilot was reported to have buzzed the main pier of the lake-picnic area, circled to the south end of the lake area and was flying northeast when the plane hit the wires. The plane was rented from Stewart's Airport. The wires are reported as carrying 6,900 volts.

1949

May 20

C. R. Martin, a Morgantown pilot, and W. F. Hickman, a passenger, were slightly injured when their plane crashed in a hollow on College Avenue side of the university dairy farm, one-half mile west of Morgantown Municipal Airport.

Upon investigation it was found that there was dirt in the carburetor of sufficient size to stop the flow of gasoline. It is unknown how foreign matter got there.

May 29

Robert L. Shell, pilot, and Bernard Shell and Warren Roberts, both passengers from Logan, were uninjured when their Stinson Voyager N97066 crashed on the mountain above Talpin Airport. There was a low ceiling and foggy at about one thousand feet. The pilot could not see the clear sky. The pilot claimed he was climbing away from the airport and all of a sudden he saw trees. He was making a climbing turn to the right and struck the trees. The airplane was completely demolished and burned. There were no injuries.

June 2

A young ex-Marine from Dover, Ohio, escaped injury when a heavy rainstorm forced his light plane to the ground near Malden.

Hammond said he attempted to set his plane down in a field just off U.S. 60 when a storm threatened to send his plane out of control. The plane upended on its nose when the landing gear caught in three-foot high grass.

June 12

Paul H. Duffy and D. B. Evans, both of Richwood, were practicing landings in a Fairchild M62-A3 NC 69424 at Marlinton Airport. The airplane took off from the airport. During bad weather with landing flaps down, it climbed to three hundred feet and after retracting landing flaps, went into a spin and crashed into the Greenbrier River.

Leonard Dobbs, Jr., student pilot of Beckley, was uninjured while flying as a student during solo work in a Piper J-3. He crashed at Beckley-Mount Hope Airport. Dobbs stated, "I was in the traffic pattern when I noticed a storm approaching. I turned on final and made a normal landing. Then all of a sudden a gust of wind and rain hit me. I saw I was heading for the airplanes on the line and I attempted to ground loop, but to no avail. I applied throttle, jumped the airplanes and the storm carried me over into the trees."

The right and left wings, right and left landing gears were damaged and the belly of the airplane was badly damaged. From the rear seat forward, most of the damage was done. There were no injuries.

June 14

A United States Army Bomber B-25 crashed into the top of Spruce Knob in Pendleton County, approximately 100 feet down from the top of the east side. The elevation of Spruce Knob is 4,860 feet—the highest point in West Virginia. Four officers and five enlisted men were killed. All were from the state of Utah. The cause of the crash is believed to have been bad weather.

June 19

Floyd Chapman was uninjured when his plane hit a tree after developing engine trouble. The accident occurred one mile west of Princeton. Agent John Gibson investigated the accident.

July 24

W. K. McDonald of Pennsylvania was uninjured when his BT-13 ground-looped in an attempt to takeoff from Philippi Airport. The pilot said he noticed that it took an unusual amount of rudder to hold the airplane straight after opening the throttle for takeoff. He throttled back and continued to roll down the runway about three hundred feet and the right wheel came off causing the plane to ground-loop and nose over, bending the bottom cowling and one end of the prop. Apparently the cotter pin on the right wheel had sheared off and the bolt came off causing the right wheel to come off on the takeoff run.

August 6

C. L. Hughart was uninjured when he swam away from his plane after having landed in the Kanawha River. The pilot had been on a cross-country flight to Glen Dale and returned to Kanawha Airport.

He then flew down to Clark Field where he made a landing, taxied to the lower end of the airport and took off. Three-fourths of the way up the field, the engine quit. He landed in the river. There were no injuries.

October 18

H. C. Beverage's Clarksburg private pilot certificate was suspended for six months from the date October 18, for low flying in the vicinity of Fairmont, West Virginia. While circling a baseball field, he struck a power line, but was able to return to the airport and make a safe landing.

November 6

E. R. Butler of Charleston, while returning from a cross-country flight from Charleston to Parkersburg, landed a seaplane in a flat bottom field five miles east of Webster Springs. The visibility was eight miles, wind gusty at ten miles per hour, and the ceiling unlimited.

CAR GAS IN AIRPLANES

The beautiful little island I want to tell you about has disappeared. This island was only a hundred feet below the Montgomery Bridge. A coal company has laid a heavy hand and filled in the waterway between the island and the mainland, making it one enormous slag pile.

July 1950, I was to fly to Charleston to pick up a passenger, only to discover I was out of gas. I hastened to the nearest gas station and purchased ten-gallon of car gas and then departed to Charleston. As I approached the Montgomery area the airplane engine acted up, making it impossible to continue. The little island lay directly ahead. It had been used as an airport some years before. I had no choice, but now there was a small growth of locust trees covering the island. By using the hand primer I could get small spurts of power from the engine allowing me to make the island and settle the Cub among the locust trees, being the same height as the airplane, with no damage to the airplane.

There was a submerged wooden bridge and using the bridge, the water was only waist deep. I waded across, reaching the mainland. I headed to the nearest state police headquarters. I was dripping wet and they were not interested and would not allow me to use their telephone. Finding a phone at a nearby store, I called the airport for help.

My BSA Explore Unit responded in full. With axes and saws, chopping for hours, we cleared a runway. They brought me fresh gas. The trip I started early that morning ended late the same evening, but without my passenger. Don't use car gas in airplanes.

THEY WILL FLY WITHOUT YOU

On September 2, 1950, George Spade, pilot from Harper, West Virginia, encountered poor visibility and seeing gas running low. So he made a successful emergency landing near Peterstown, West Virginia.

In order to tie down the plane for the night, he attempted to move it to another part of the field. He proceeded to start the airplane by hand and without help or chocks beneath the wheels and no one to hold the brakes, the little bird departed on its own. It ran into a tree, doing extensive damage.

SOUTH OF THE BORDER
1950

The Pownee Piper sprayer airplane circled the airport. It was the first landing after leaving the factory at Lockhaven, Pennsylvania.

Stepping from the airplane was a fine-looking young man, bewildered, confused, and scared. The reason . . . he was from South America and he could not speak English. He handed us a Standard Oil Credit Card. The story—the amount of gas his airplane had used told the hours he had been in the air.

Each person at the airport contributed sandwiches, fruit, and candy to help him with a safe journey. As he departed, the expression was that of gratitude as he waved good-by and headed South.

LOW-RECKLESS FLYING COSTS PILOT

On June 18, 1949, Francis C. Holcomb, twenty-three years old, Alderson, and Richard E. Alderson, age twenty-four, Alderson, were slightly injured when the plane they were flying crashed into the Greenbrier River. Witnesses state that the crash was due to low and reckless driving. Sgt. S. C. Ballard and Trooper D. A. Honaker investigated the accident and both Holcomb and Alderson were directed to appear before Justice of the Peace Finley Arbuckle in Lewisburg. On June 27, 1949, Holcomb and Alderson appeared before the justice as directed and Holcomb took upon himself the full responsibility for flying the aircraft and entered a plea of guilty to operating an aircraft in a careless and reckless manner in willful or wanton disregard of the rights and safety of others or without due caution and in a manner so as to endanger or be likely to endanger any person or property and was fined two hundred dollars and cost and released. Charges against Alderson, who was a passenger, were dismissed. The aircraft was rented from James A. Tolley of the Hinton-Alderson Airport, Pence Springs, West Virginia.

HAIL
1952

It was one of those hot July days and there was much flying. As each student finished his flight, he tied the airplane down.

There was no thought of a storm. I went to town on business, only to be sent for. A hailstorm had destroyed five airplanes and the roof of some of the hangars. The stones were as big as golf balls, and the storm had made a narrow path across the airport about two hundred feet wide. The hail lying on the ground was several inches thick. It could have been shoveled up. The warm sun soon melted it, leaving only evidence of its passing through.

Visitors who had not before or since been on Fayette Airport, came sight-seeing.

The next morning I was awakened by the sound of hammers. My brother and others were putting roofs on the buildings. This was a time when you knew who your friends were.

1956

As a Tri-Pacer, piloted by H. A. "Buddy" Sessler, was rolling for takeoff at Fayette Airport, the entire fabric covering of the cabin fuselage top gave way and peeled back, covering the entire tail assembly. This occurred just as the plane was lifting from the ground. There was no elevator or rudder control. Buddy Sessler immediately cut the power and landed safely. After gluing the fabric back, he returned to Princeton.

The fabric on the top side of the cabin is a weakness. This has caused several accidents.

FOR THE RECORD

The aeronautics commission has been buffeted by internal strife on several occasions since Governor Cecil H. Underwood, about midway through his term, succeeded in having an aviation pioneer, Hubert Stark, fired.

C. Steve Hanifin of Summers County replaced Stark. Hanifan lasted until Underwood left office and W. W. Barron became governor. Barron's first appointee was James A. Downs of Morgantown, who was opposed by some of Barron's closest political associates, and he was replaced.

Parrish was the choice of the group for the small-pay job, and he lasted through the remainder of the Barron administration and well into that of Governor Hulett C. Smith.

Smith first came into state government as a member of the aeronautics commission.

Governor Smith then replaced Parrish with John A. Wilson, formerly of Oak Hill. Governor Arch A. Moore, Jr., brought with him his own personal pilot, Floyd Graham, to fill the seat of commissioner. Graham drowned in New River, Fayette County, and was replaced by William E. Richards.

1958

On January 22, pilot and passenger from Morgantown, West Virginia, were fatally injured when their light plane crashed on New Creek Mountain, fifteen miles from Keyser. The cause is believed to have been bad weather and low ceilings.

On January 28, a twin-engine Cessna from Clearwater, Florida, en route to Charleston, encountered low ceilings and fog, and crashed near Bradshaw, West Virginia. All five aboard were killed.

On January 31, Air Force C-45, en route to Charleston from Florida on a training flight, crashed into a mountain peak eight miles from the Charleston Airport, due to fog. The accident fatally injured both occupants.

The three accidents mentioned above happened within a period of nine days. All were bad weather accidents.

AMBUSH
1961

Pilots fly in where policemen dare not tread.

Pilot Dennis Workman was flying the state of Oak Wilt Blight Survey of the plateau area of the upper watersheds of Panther Creek in McDowell County.

A low flight was in progress for the purpose of identifying a blighted tree, when, from the ground, a shot was fired from a heavy caliber shotgun. It damaged the airplane and barely missed the pilot and observer.

This was immediately reported to the Welch Department of the State Police. After much discussion between pilot, police, and politicians, it was decided if there were any investigation that it may disturb every moonshiner and bootlegger in the area, making it unsafe for any flying in this location.

APRIL 27, 1963

Federal Aviation Agency
General Safety District Office
Municipal Airport
Charlotte, North Carolina

I wish to submit this letter as the best explanation I can give for declaring an emergency near Hickory, North Carolina.

At 12:10 EST I had completed filing a flight plan to Fayetteville, West Virginia, altitude 3,500 feet, climbing to 5,500 feet. The plane without warning, in a combination roll and over the top spin commenced a wide slow uncontrollable spin, the controls were without pressure as though the tail assembly was off. At a low altitude the spin broke, the plane was then recovered as though it had been in a complete power stall. It was my impression, at the time, that we had lost a large portion of fabric from the fuselage which later proved to be false.

Flight Service Station located us and directed us to Hickory, North Carolina. It was the opinion of Flight Service that I had gotten in a Prop Wash of a Piedmont Airline in the area at the time.

The plane was checked and flight checked and returned to Fayetteville, West Virginia.

I have no further information other than the weather was

good and I had been in the air 3:45 minutes, having departed from La Grange, Georgia, and my left gas tank was almost empty at the time.
 If I can be of further help please call on me. I am
 Respectfully yours,
 Frank K. Thomas
 7000 hrs. plus
 Instructor, Commercial and Instrument

FAYETTE AIRPORT

During the time we were paving the airport we placed a sign which read as follows:
 "This project is not your tax dollars spent or wasted. Free Enterprize at work."

40 WEST

Herbert Alloy of Beckley was very much concerned as to the whereabouts of one of his students. His voice showed much concern over the unicom radio as the weather deteriorated. Finally, after many calls, the student replied, "I am about forty feet." Alloy quickly asked, "Forty in which direction?" His answer, "Forty feet west of the gas pump."

THIS YOU SHOULD KNOW

The jet blast of a Wright J-65 engine turning up at full power has a temperature of 380° F. and a velocity of 271 K 25 feet AFT of the tail. At 50 feet the temperature is 213 and the velocity is 102; 300 AFT should be clear. Sometimes in the air the wake turbulance lasts for fifteen to twenty minutes.
 On November 11, 1972, McCoy Air Force Base, Orlando, Florida, an FBI agent was blown over one hundred feet and his clothes torn off, from the blast of a twin Jet DC-9.
 This was being hijacked.

FROM THE FILES
Charles Yeager

Charles Yeager, the famous pilot from Hamlin, West Virginia, thrilled, chilled, even angered many, when he put on an aerial show over the city of Charleston at speeds approaching the speed of sound.

The story has the speed at 700 mph when he flew directly under the South Side Bridge.

FROM THE FILES

The morning of December 7, 1941, the United States Army airplanes available were 3,000 and only 1,157 were suited for combat.

In a matter of hours on December 7, 1941, the Japanese reduced our combat suited aircrafts to 800.

1951

Kelly Skaggs, Merle Grimes, and myself were en route to Tulsa, Oklahoma, when we encountered a line squall of intensive thunderstorm action with a lot of lightning. We flew a lengthy amount of time northwest to find a weak cell. Upon finding such, we lowered our altitude and started through the hole that appeared to have some light shining through. Kelly asked, "What's causing the engine to back-fire?" Checking all possible causes, we discovered we were over a canyon firing range. It was cannon fire we were hearing.

I might mention that we would nearly jump out of our seats when the whistle sounded on the old train one thousand feet below.

WHY NOT UNDER THE BRIDGE?
1950

Clarence E. Searls of Kanawha County was fined twenty-five dollars and cost on May 30, 1950, on a charge that he unlawfully operated an aircraft, Luscombe N219953, at an altitude of less than one thousand feet and did buzz the city of Saint Albans.

LEGAL AT LAST

Pete Puckett told the story about when he was on his first legal instrument cross-country. How proud he was. He considered himself now in the big league, legally at last. It was evident that in the past thirty years he had done a little scud and illegal flying. This time he was piloting for a large company and the president of the company was with him. He then stated, "Everything would be fine if I could find time to get this aircraft licensed. The annual has been out for one year."

'LOST' NAVY PLANES CIRCLE OVER BECKLEY

This article was taken from the *Raleigh Register*, January 31, 1954. This could not happen now.

> Fifteen Navy planes circled over Beckley early Saturday night in the belief they were over Charleston.
>
> Raleigh County Memorial Airport Manager C. Harold Hanks notified the air traffic control tower, Charleston, that the planes were over Beckley. The planes were directed to Kanawha Airport, where they landed for a fueling stop, after the Charleston airport checked their progress by radar.
>
> The control tower operator said the group included 14 Corsair-type fighters and a C-54 lead plane. One of the planes had temporary engine trouble, but landed in Charleston safely, the control tower operator said.
>
> The flight originated near Norfolk, Va., and was believed en route to Columbus, Ohio, or Minneapolis.

The lead plane flew ahead of the Corsairs and arrived over Charleston, the operator said, but the others apparently circled Beckley in the belief that they were over the capital city.

The crews planned to stay in Charleston Saturday night, and depart at 8 a.m. Sunday.

OCTOBER 27, 1973

Chester Seaman, Jr., returned home from training in the Air National Guard. He was a skillful pilot with a recently issued commercial license, with a twin-engine rating.

Chester, being a former student, came to me for help. It was necessary he be recommended for a single-engine rating to be legal. After the necessary flight time, his recommendation was carefully prepared. After presenting this to an examiner, it was rejected because it was not typed. It was neatly printed. Complying with the examiner's wishes, it was typed and accepted.

Upon the successful completion of the test, he was presented with a scrawled, printed, temporary permit. It was handwritten with mistakes.

P.S. Is there a Civil Air Regulation calling for a form to be typed?

TIMING

Dr. James K. Pickens visited Fayette Airport stating his plans, while here on his vacation, were to obtain his private pilot's license. He explained the time and money he had for the flying course.

Things went on schedule as he received his license. He stated, "It certainly is strange how my money and time worked out to the exact penny and minute."

FLIGHT BREAKFAST

Until just a few years ago, many of the state airports held an annual flight breakfast, being host to all pilots who wished to attend.

This breakfast was held in the hangars or out in the open air. The food was usually free to pilots and passengers. There were prizes given to the oldest and youngest pilot and also the first arrival and the one who came the farthest distance. It was a wonderful homecoming filled with great fellowship.

FLYING PRIEST MAKES ERROR
Lands on Long-closed Strip, November 1967

A Welch priest, who has been flying since 1952, landed his plane on the old Princeton Airport—and was he embarrassed. The airport has been closed three years.

The Reverend Father Karol R. Radzieta, fifty-six years old, said he was in contact with the Mercer County Airport, and not being familiar with the area, thought the Princeton open space was the Mercer airport.

The priest flew in Saturday from Somerset, Kentucky. Onlookers said he circled the old airport before landing, and was doing very well until the craft ran into soft dirt and stopped abruptly.

State police and Princeton city police helped get the Reverend Father Radzieta off the ground and headed toward Welch. The Welch Municipal Airport is about thirty miles away.

FAA agent, Paul French, discussing the mistakes of the lost priest at Charleston before a gathering of forty flight instructors, was a little critical until a stately old gentleman, Joe Reeding, asked if there was one instructor out of the forty which had never been lost. There was no answer.

FAYETTE AIRPORT
1968

Harvey W. Summerfield of the Charleston police force was practicing takeoffs and landings. As he was commencing a go-around his engine sputtered and lost power. When he released the flaps, the engine gained some power allowing him to make a successful forced landing. The cause was found to be an insect, probably a mud dauber, which had set up housekeeping in the gas tank vent, stopping the flow of gas to the carburetor.

NEWSPAPER

I have long wished to challenge the right of Beckley newspapers for the little respect they have had for the deceased and the families of the victims of air crashes. Never have I seen such in other newspapers; the close up pictures of those killed in airplane accidents. Perhaps they have never stopped to think what this does to the families of the deceased. Do they care?
No one cares for them showing the pictures of the wrecks. If the wreck was careless it would serve as a warning to all of us. But what does the newspapers gain by showing the battered bodies of our friends?
I offer the heartiest congratulations to the state police and employees at Bolt Mountain for keeping those who were not part of the rescue party off the mountain during a recent crash.

LOST PLANES

The saying is that the pilot who has never been lost is one that has never been anywhere. Ninety-eight percent have been lost, the other 2 percent are liars.
In 1943, an army plane was lost and landed at Huntington, West Virginia. One of the crew crept from the airplane asking

an airport attendant, "Where are we?" The answer, "Huntington." He turned to another of the flight crew saying, "I bet with you we were in Indiana. Pay me."

In 1946, an army cargo plane was lost en route from Washington to Cincinnati. Bad weather had exhausted their fuel supply. Pilot George Neffinger from Chicago was unfamiliar with the mountain terrain. They were making preparations to parachute out when they saw Fayette Airport and they then landed safely. Fayette Airport was then one-half the present size, so some fences were removed to insure a safe takeoff.

On November 1, 1965, a Piper Colt airplane landed on the Oak Hill expressway. Piloted by Dave Groves and accompanied by his wife, Mrs. Groves, they were en route from Huntington to Pittsburgh. They had followed the wrong river and were totally out of gas, so they landed safely on the expressway. The plane was flown off the next morning by Thomas to Fayette Airport where it was refueled. The pilot and passenger continued on their way.

HEAVEN OR HELL

This would be my interpretation of hell. There would be a field full of PA-18 Cubs that could not be worn out. One would be sentenced through eternity to fly Forest Oak Wilt Flying. This would be hell.

Heaven would be a field full of 150 Cessnas with all those young people who did not have a chance to fly on earth. My assignment would be to teach them throughout eternity. This would be Heaven.

THE REVEREND W. JONES

Charleston Approach, this is Cessna 547 over Montgomery, thirty-five hundred feet inbound landing Charleston. AMEN.

FOR WHAT?
1948

Frank Rosscano, flight instructor at Fayette Airport during the early days of its operations, saved the office from burning when he discovered the stove pipe had fallen from the Burnside coal heater. He rushed to the nearest house and carried several buckets of water extinguishing the blaze.

Upon my return he was very proud of his accomplishment. Then I asked him, "Why did you not use the fire extinguishers?" He replied, "I thought you were saving them."

For what?

POWER LINES AND LOW FLYING

With all the criticism I have given toward buzzing with the tragedy that follows, there have been power line incidences—some careless, some unavoidable.

W. F. Bays, a Fayette flying product, one of the nation's leading agriculture crop dusters working in the Midwest, tells us it is quite common for the duster airplanes to deal with power lines.

The crop duster, flying as low as possible, sees a power line that he was unaware of and too late to pull up over it, puts full power to his airplane and cuts the line, usually without damage to the airplane.

Bays tells us that they are not permitted to use sunglasses while flying crop dusting. Glare is part of vision and the dark glasses cut down the vision.

The experiences between our boys and girls and power lines have been numerous.

Mrs. Arbury Light, while landing her Super Cruiser at Boone Field in 1948, caught a downdraft, raking the left wing across a very live high voltage power line, burning small holes in the wing. Mrs. Light skillfully landed the airplane without further damage.

To follow, a few days later Richard Bennett removed the power lines from the end of the old Summersville Airport without damage to the airplane. This was just at darkness with the same make and model Mrs. Light used.

Two boys flying from the Jones Flying Service, Columbus, Ohio, were lost. They attempted to set down on the highway, at the Oakridge Addition, Fayetteville. They straddled the power line, easing the airplane gently to the earth, without the slightest injury to pilot or passenger.

My personal experience with power lines are explained in "Miracle or Coincidence." But there was one other at Craigsville Airport. The wind was very strong and I saw I could not climb over the top. The airplane was too small to cut the line so I went under it.

Another experience was when the power company decided to put a power line at the end of my airport. They put the stakes in the ground and someone removed them. They came with their poles and I came with my ax. The poles were never put up.

All low flying airplanes must respect the turkey raisers. The turkeys, sometimes over one thousand in a flock, panic at the sight of an airplane, running to the corner of a fence and bunching up, smothering. This causes much loss to their owner.

Those that raise minks experience loss, too. The low flying airplanes panic the minks and they kill their young.

I was once hired by the United States Department of Wildlife for a flying job to count all the wild geese and ducks on the lakes and creeks in West Virginia. This was a week's job—a real wild goose chase. I suppose one of the bureaucrats wanted a long airplane ride. We may have counted the same goose several times, but there was one that we forgot to count—the one that hired me.

From my files I find that on September 29, 1958, two 5/16-inch galvanized steel wires were snapped cleanly in two. This happened by a navy jet near Bluestone Dam. Comments were that the accident probably kept the pilot from crashing into the hillside. When the plane hit the power line the pilot realized that he was flying too low. Oddly enough, the jet suffered only minor damage. Its radar dome was broken and the rudder was dented, with the canopy severely scratched.

My personal comment is that we all know our own freedom of today has come from the military flyers protecting our country. There is no such thing as traffic control as long as military planes are allowed to fly as they please.

When I made complaints to the FAA they said that these flights were only made over isolated, scarcely populated areas. What of the youth, whose mother waits for his return, in a light plane and encounters wake turbulence and downed. I spent five months in a wheelchair because of these low flying jets.

LOW FLYING

Eugene Fourney, with a rented plane from Princeton, was showing his low flying skill to the girls at Concord College, Athens. Not only was he too low, he was at the wrong place at the wrong time. John Gibson, FAA inspector, was on the ground at Athens. Needless to say what happened to his license.

WIVES

Case I

Rosco was ordered to sell his airplane or get himself another wife. He told her that wives were easier to get than airplanes. His new wife is a lovely person.

Case II

Pronto's wife told him if he bought a new airplane she would drive his car into the river. He did; so did she.

DUST

A lady called Fayette Airport complaining of all the dust blowing her way each time an airplane takes off. The lady asked us what we were going to do about it. I replied, "I am trying to blow it all away."

MEN

I asked myself, "Am I in the wrong rest room?" when I stepped up to the sink and found it full of long, blond hairs. I rushed out thinking that I was in the wrong room. With a sigh of relief, it did say, men, this was Morgantown, West Virginia.

WORK

A young man asked if he could do some work for an airplane ride. "Yes," I replied, "there is a broom or a lawnmower." He said he wanted to ride first and work later. I told him that would be all right with me because I might be busy later.

After landing, I pointed to the lawnmower or the broom. He demanded, "Take me home. You made me air sick." I replied, "Get to work." His reply, "I have no work permit. I am only fourteen. You will get in trouble if you work me." I then told him not to make the mistake of coming back when he turned age twenty-one.

MORGANTOWN

In Morgantown, West Virginia, while waiting for takeoff clearing, there was an Allegheny airplane on final approach. The tower informed the Allegheny airplane that there was a large bird circling at the end of the runway. It was either a hawk or a buzzard. The pilot of the airplane quickly replied, "Is he instrument or visual flight rules."

NEVER MAKE IT

While on an air ambulance trip, my ailing passenger told me he did not think he was going to make it. I told him to relax and not to give up because we would be landing in minutes and there would be an ambulance waiting which would rush him to a nearby hospital where he could get special treatment. He again replied, "I'll never make it."

Then I realized that it was a bedpan which he needed and we readily supplied it.

TRUE HUMOR
1950

Dr. Douglas Freeman was the historian and director of Southern Railways. While addressing a meeting of top officials and six hundred employees of the Norfolk and Western Railroad at Roanoke, attacked, and rightfully so, the subsidiary received of the airlines by the government. He presented statistics to show that trains are safer than planes.

Then, after his speech was completed, he excused himself saying, "I will be late catching my plane home."

FLYING BUG

When bitten by the flying bug, there is no known cure. For temporary relief, visit your nearest flight instructor for continuous treatment by flying until joy is obtained. Warning! If flying is not continued, some agony will reoccur.

THE WORST OF ALL

It is of no apology that I am against alcohol or strong drink, and do not permit it on Fayette Airport.

We do not need those weak individuals who get their courage from a bottle or a glass.

Those who permit bars to be set up at airports are sacrificing safety and good common sense. Are they so weak that they cannot make a living from as profitable a business as aviation, or are they so hoggish after the dollar. This is also true with the airlines which have made themselves flying barrooms and the young beautiful air hostesses barmaids.

Is there one person that will say that if the whiskey is on board an airplane among all of our pilots that there is not one unknown alcoholic?

I personally signed a complaint against a major airline selling alcohol beverage illegally on the ground and in the air over the state of West Virginia.

With all the efforts we have put forth against alcohol, upon two occasions at Fayette Airport, pilots have been caught flying and drinking. They fly no more.

WIND—1946

March 11, 1946, was the clearest, warmest, most beautiful day I had experienced for that particular time of the year. As evening came, the barometric pressure was falling. There was no doubt that bad weather would soon be with us. As night came, there was a thunderstorm with lightning and followed by snow. Later the wind blew, increasing in velocity. By 2:00 a.m. the barometer was still falling with winds 50 mph with the worst yet to come. I had turned the airplanes with the tails into the wind.

There was only one building on the airport, a large trailer. It shook as though it would turn over. Seventy mile-per-hour wind gusts were hitting it. The wind was tugging at the ropes of the airplanes. It was obvious that they would not be there long. For further precaution, I dug holes in the soft ground putting the wheels of the airplanes in the holes, and then filled the airplanes full of cinder blocks for additional weight. I then pulled the barbed wire off the fence for additional tie-down.

The next morning the wind had subsided. Our airplanes set guilty, frozen in the holes. I soon learned that Hilltop Bolinger

Airport, Charleston, had lost sixteen airplanes. The fittings of the airplane broke loose. Also, Beckley and Pence Springs each lost two.

WIND OF 1968

Darrell Kidd of Beckley was a student pilot at the Fayette Airport. Darrell was asked if he wanted to fly. "Yes," he replied, "ride around the field once with me. I have not flown lately."

The air was calm and clear, not a cloud in sight. The aircraft, a PA-11 Cub, checked out perfectly. This was one of the days that an instructor enjoys sitting on the ground and watching the student's progress.

The takeoff was normal. Then in an instant, a wind hit us that slowed the airplane's ground speed down to the extent where we would have ordinarily been fifty to one hundred feet. We were one thousand feet. I told Darrell to circle around and we would terminate the lesson. Following the usual pattern around the field on base leg the wind blew the little Cub so far from the turning point that if the actual distance was told the story would seem too fantastic to believe. So we will omit that. "It was just a weather phenomenon. It will pass as quickly as it came," I told Darrell. We gained altitude and started a high approach going for another try, finding that the Cub actually backed up. We then decided to sit it out in the air for a while. One hour and fifty minutes passed. We had no radio and the airplane engine became sick and lost power. This was it. We had watched people come and go in their cars at the airport and none seemed to know our trouble, high wind, bad engine, low on gas and extreme turbulence.

We were not sure of the wind velocity but we knew that it had let up some. It was now or never to land. The last two hundred feet from the fence we were below the fence level. Usually at this point there is a downdraft and we hoped that the engine would give us an additional shot of power. As we reached the fence I thought we would hit it or straddle it. Our prayers were answered. As we reached the fence below the level of the fence, I pulled the nose of the airplane up to ease the impact on the airplane as it hit. Over the fence we went, clearing it by inches and landing on the overrun area from the runway, which is downgrade from the runway.

Those land observers thought that we had landed there to shelter us from the wind. There was no damage and we found that the trouble was in the carburetor.

WIND OF 1947

Paul Neal was flying with W. F. "Red" Bays. The wind increased. I had decided to terminate all flying until a quieter time.

We watched a Cub approach, and as it started to land, the wind picked it up and flipped it up on its back. We all rushed to the scene, helping and instructing the occupants out of the airplane to prevent damage.

My feelings were so hurt that I was not noticing who I was helping from the airplane. Looking up, I saw Glen Holley, an employee of my competition. As I began to ask him where he had come from, I realized that it was he and one of his students that had put the airplane on its back. This was not my airplane.

Paul Neal and Bays landed shortly without incident.

TORNADO
June 23, 1944

Most West Virginians have forgotten the worst tornado ever to strike the United States of America. The deadly twister hit Shinnston, West Virginia, at 8:30 a.m. on June 23, 1944, killing 103 persons and injuring hundreds of others. Six hundred fifty homes were destroyed and 411 damaged. That same evening, 150 people were killed in West Virginia, Maryland, and Pennsylvania, and it was believed to be from the same storm.

I had never witnessed a tornado before, but the night of June 23, I was working at Alloy and I actually saw a fifty-five-gallon barrel blown out of sight.

STRONGEST WINDS

From the records of the West Virginia Weather Bureau the following records were found:

On April 28, 1966, winds from the west at 28 mph sustaining winds exceeding one minute in length with gusts to 48 mph; and in 1967 on two dates, February 16, winds at 37 mph with gusts to 59 mph, and May 7, winds at 32 mph with gusts to 61 mph.

The all-time record was set on May 31, 1953, with winds from the west, southwest, wind at 55 mph with gusts to 76 mph.

CARSON GREEN
1953

Carson Green was with me on a trip to Florida. As darkness came, the radios ceased to work with all the electrical system making navigation difficult. Carson was flying, and when he made a sharp turn, I asked, "What are you doing?" He replied, "I am going to follow that jet." The jet's speed was 450 knots and our tri-pacer was 100 knots. My reply was, "Not with my plane." Soon we saw far over the horizon a green and white beacon and we then knew it was a lighted airport.

GAS THIEVES
1948

Each of the employees were given a key to the gas pump. This way business continued when duty called me elsewhere. The pump master figures were checked daily. But at night more gas than one car could hold was taken from the pump. Only someone who had knowledge of my going and coming could be guilty, and it had to be an employee.

Each evening I switched the lock and then would switch back at daylight. One morning one of the employees told me he needed a new key, that his did not fit. I asked him to go with me and try it. This time it worked. He was told then that only at night his key did not work. He was discharged. He admitted taking gas for himself and a friend.

WIND

As we explore the elements, first we will find that one of the causes for wind is that as the hot air rises it is replaced by cooler air moving in.

Speaking of hot air going up, it is clearly understood by the hot air and wind in the following.

DESPERATE FOR OFFICE
More Wind—Sunday, May 7, 1972

Pat Hamilton, candidate, released a newsletter of which we take contents. It is as follows:

> I regret that I am so vehement opposed by the Fayetteville 400 Social Set, some of the officials connected with and on the Fayette County Board of Education, and Frank Thomas, who operates what he erroneously calls an airport. I am better off with them against me. I stand with the Blacks.

Thanks, Pat. Your article gave us a good day of passenger riders. It was free publicity.

WIND
1949

As the wind rattled my windows, 2:30 a.m., it was time to rise from my slumber, leave my bunk, and go to the airport to check wind damage. Those airplanes that were tied down were dancing, rocking and tugging against their ropes, wanting to be gone with the wind. All were checked, blocked, and ropes tightened.

Now I had to check the hangars to see if the wind had worked any doors loose or pulled any tin loose that may have to be nailed down.

Upon reaching Lee Frazier's hangar (this building had not yet had doors installed), I found that the wind had picked up the entire hangar building. The building was moved eight feet back without damage and the airplane tied down in the building had suffered no damage.

I rushed to the telephone, calling Lee Frazier. I told him, "Your hangar is blowing away; come and help me move your airplane to the big hangar." All I heard was a click of the receiver. By it being 2:30 in the morning, I supposed he thought I was joking. I called again. This time Mrs. Frazier answered. Her reply was, "What's going on, Frank?"

Lee grabbed his boys, ran out of the house, and within minutes he and his two sons arrived, clad in their pajamas, and the airplane was moved without damage.

The hangar was put together so well that it suffered no damage at all. But, I assure you, the next time it was bolted into concrete, where it has remained in use protecting airplanes for the past twenty-five years.

WIND
1970

The situation repeated—this time during daylight hours. The wind was gusting 65 mph. A check on planes and buildings revealed that the wind was roughing up the judge's hangar and soon would destroy it.

I called the courthouse for the judge but he could not be reached. The word was out that we needed help and it came fast. One of the boys said that he would go for the key to the airplane, but we all agreed that time would not permit it. By hand, we removed the airplane to a safer building. We were not more than minutes out of the building when the wind raised the roof straight up with a one-thousand-pound steel beam attached, then crashed to the ground. The building was nearly destroyed and minutes earlier, would have caught at least six men in the building.

THUNDERSTORM
1946

The date, 1946; the time, 7:30 a.m.; the temperature, seventy-five degrees; visibility, very poor; and the airplane was a J-3-C Cub Tandem Trainer. The instructor always rode in the front seat. This was the custom in case of an accident, causing the instructor to receive most of the impact, seldom hurting the student in the back seat. The gas tank was strapped in the cabin, just above the instructor's legs, practically in his lap. The gas gauge was a cork with a wire through it, the cork pushing the wire through the gas cap causing the wire to drop from sight when the gas tank became empty.

It is worthy to note that in those days the government frowned on side by side airplanes and in some programs forbade them for instruction. They reasoned then that the student would only watch the instructor instead of reacting for himself. Today their thoughts have reversed.

The purpose of this particular day's flight was to practice takeoff and landing with a student, Harold Owens. Usually, this particular time of the morning was an ideal time for a lesson, after a restful night for both instructor and student, with no spectators to excite the cause. But not this time.

On our second approach for a landing there came a thunderstorm from out of the haze and one mile visibility, with movement so fast, winds so strong, and carrying with it such low fog and stratus clouds, that landing was just impossible. We were forced to flee the area. Never before, or since, have I seen a storm with such magnitude. As we were flying East, it appeared that some of the storm was getting ahead of us.

The little wire that told the story of the gas was barely showing. We would have to do something soon.

Each strip mine we would pass, Harold would point it out to me. I would tell him that there might be something better a few miles farther. We were in a sparsely settled farming section and before my very eyes I spotted a field that appeared to be large enough.

There was really no choice with the storm coming so fast and the gas gauge showing empty. This was for us. With a terrific wind, there would be no trouble landing. Directly ahead of us was a very large barn. We taxied the plane to the back side where we were partly sheltered from the storm. As the strong winds reached us, we stood beside the airplane holding it to prevent damage from the wind. We actually had a small shock from a bolt of lightning that struck nearby. The storm ended and all was well.

A farm lady came from the house, very surprised to find an aircraft in her backyard. With a very loud voice, she yelled, "Hey, Paw. Come here and see what's in our backyard. It's an airplane."

We soon learned that we had landed at Corliss, West Virginia. The power lines and telephones were down due to the winds from the storm, and we could not even get car gas because the pumps were not working due to lack of electric power.

Back home at Fayette Airport, the search had begun for us. Charles Haga and W. F. Bays had taken to the air.

I rented a truck, drove to Rainelle and called home for gas to be brought.

When it was time to take off, we found that the field had shrunk. It was not as large as we had thought. It was only three hundred feet in length.

Leaving Harold to return by car, to lessen the weight in the airplane, touching the red brush on my takeoff, I was airborne—Homeward Bound.

GAS CRISIS
December 7, 1973

As duty called a flight to be made to Dulles International Airport—an air ambulance trip—we had our first experience of fuel crisis shortages. Upon landing, we were promptly asked for a three-dollar landing fee and told that they only had gas for the airplanes based at Dulles. As I waited for the ambulance to arrive, I realized that with luck we would have gas to make it home, but darkness would be on us. But if the weather turned bad, we would have no gas for an alternate airport.

While in their office, a call came over their aircraft radio that Jet N1620 would like three-hundred-gallon of gas. Their gas truck met N1620. And as we watched, all that was aboard that jet was the crew and two passengers, which got off at Dulles. A government limousine pulled up beside the jet and picked up the passengers.

I returned to the office and asked why an air ambulance could not get twenty gallons of gas, but a jet could get three hundred gallons.

I told them I was calling NBC or CBS to give them the story, and I was promptly sold gas.

GAS THIEVES
January 4, 1974

Weeks had passed and my bones ached from the loss of sleep. Gas thieves were visiting the airport at least three times a week, taking gas as they pleased. With the thought this must be stopped, I remembered at River Side Airport below Montgomery how Premier Delporto's airplane was burned by gas thieves.

By the request of the prosecuting attorney, the West Virginia State Police visited Fayette Airport and informed us that if we caught the thieves to call and they would come to the scene, and get the thieves, day or night. I told them that if I caught them they would not need the police—a doctor or undertaker would be needed.

The morning of January 4, Jim Miller and myself were waiting for daybreak so that we could fly. There was a heavy frost or light snow which revealed car tracks leading to the gas pump. Only a few days before thieves had sheared every pin on the gas pump, creating many hours of labor to repair. Tracing the car tracks, we found that the thieves had lost their gas cap, the cap being light blue in color with foreign words stamped inside. After taking the cap to filling stations and garages, we knew we were looking for a small foreign compact, blue in color, without a gas cap.

Late the same afternoon I saw the car, an Opel, on the highway passing the airport. I started my pursuit but the car had vanished, the driver turning off a side road returning a different route.

Emil Jurak and my sister had solved the case. The boys had parked in a residential section and walked back to the airport looking for his gas cap. While he and his friends searched the area, Jurak and Elizabeth searched for their car. Finding it, they stopped me as I returned from my hunt. I was waiting at his car. He was uncooperative and left me no alternative but to call the police. The youth, a handsome kid from one of the best families in the community, was a user of marihuana. One of the thieves had been released from jail long enough to allow his hair to grow down on his shoulders.

This one was taken to the justice of the peace and fined ten dollars for trespassing and ten dollars for destruction of property.

LOG BOOK

I greeted one of my long time no see students with a "Hello, Smitty. How are you?" His answer, "Frankly, Mr. T., I ain't doing so good." "What's the matter?" I asked. He said, "My log book ain't signed properly." He explained in detail, "I done gone to two of them big airports. And Mr. Burgess, Beckley, and Mr. Johnson, Charleston, don't sign the log book like you do. You just sign your name and rubber stamp all the rest of the stuff."

"O.K. So I do. You ask to see Mr. Burgess' or Mr. Thompsons' first log book and you will find the same rubber stamp as yours certified their log book."

From my log I showed him the date they both soloed here.

1968

One of the Johnson boys from near Beckley accepted my offer to solo him for a sawbuck—that is one hundred dollars in mountain money.

After thirty hours of dual flying I offered him my new Kanawha Development stock if he would quit. Harrison O. Ash was the "Big Daddy" of this stock. Johnson said, "A contract is a contract." I stuck it out and soloed him.

LOST PLANE
1973

Thanksgiving Day, 1973, a group left Florida to go to the Cayman Islands, two hundred miles south of Havana. The plane was forced to land in Cuba when a navigational instrument malfunctioned.

Cuba repaired the instrument, then demanded payment of eleven thousand dollars, which they were paid.

Our comment—I'll bet with such a charge, they thought they had landed in Charleston, West Virginia.

RICH TAGLANG
November 14, 1970

An excellent football player at Marshall University missed his team's airplane as it departed for Greenville, North

Carolina, only to learn that the DC-9 Jet crashed that Saturday. Taglang, from Bethlehem, Pennsylvania, called his parents from Huntington to tell them he was safe. Coach, players, and supporters were among seventy-five aboard the DC-9. There were no survivors.

INSTRUCTION
1973

While flying with student pilot Michael Swartz, the radio ceased to work. First I jolted the radio with my hand, hoping it would start working. Mike said there was no use trying that. I explained that sometimes a push or jolt would start them again. He replied, "No use trying. If my landings did not jolt it to working, nothing short of a crash would start it."

WHO IS LOST?

On returning from a night flight from Orange, Virginia, with Mr. and Mrs. Robert Abbot and son, after one hour and thirty minutes, we arrived over the metropolitan area of Fayetteville, West Virginia. Mr. Abbot asked me, "Will you be able to find the airport?"

THOUGHT

When you are flying and there is a little trouble, your first thoughts are:
"I will land and get it fixed at the first major airport with a repair shop." Your next thought is, "I will land at any good airport where it can be flown off of." As things get worse you think, "Any place where I won't get hurt will do."

CHRISTMAS
1973

A call came to pick up a young sailor on leave from the Great Lakes Naval Center. He had hitched a ride to a place he called Franklin, Kentucky. He called his mother, Mrs. Audrey Sizemore, who in turn called me to fly to Franklin to get him. I told her I would pick her son up at the Bowling Green Airport. All was agreed upon when she called again.

The boy said he was 135 miles from Bowling Green. There was something wrong somewhere.

I told her I would not leave until he found out where he was. After several calls he declared he was at Franklin, Kentucky, and that he would wait on me at the Capitol Airport. This cleared the way to go—he was at Frankfort. Even after we picked him up he declared he was at Franklin, Kentucky.

SEPTEMBER 25, 1973

A call came from the state police that there had been a reported airplane crash near Wonderland. This was during the time of one of those flying saucer scares. The officer wanted to know if we had any airplanes missing.

The callers described its appearance as an orange glow with some scattered aluminum through the trees. They also stated that there may be someone moving about the crash. The people of the area wanted the police to investigate.

I explained this was an old story. If they had seen anything, the persons calling would have raced to be the first to the wreckage. Even those only barely able to walk would be climbing the mountain. This story has been the same hundreds of times. If there is a crash, the people rush to it, grabbing souvenirs which should not be removed. These parts stolen and carried off may be the very piece needed to pinpoint the cause of the crash.

MARCH 1974

The following was taken from the *Charlie Brown News*, Morgantown, West Virginia, March 1974. It was written by Steve Weaver, West Virginia's largest Piper dealer:

> Frank Thomas of Fayetteville popped in during a stop in Morgantown the other day. Talking to Frank is always enjoyable since he has probably personally taught more people to fly than anyone else in the state—and possibly the country—during the 30-odd years he's been instructing. You just naturally collect a lot of good stories in that amount of time.

SILVER WINGS FRATERNITY

With much pride, I'm wearing the little wings—S.W.—Silver Wings. Only those who have stuck it out for twenty-five years of flying are entitled to be a member of the Silver Wings Fraternity, Box 1228, Harrisburg, Pennsylvania.

TWINS

The twin teen-age boys arrived at Fayette Airport almost dragging Daddy Kirkwood in, hoping for some encouragement. They had found an old war surplus airplane and the gas consumption was one hundred gallons per hour. They had soloed J-3 Cubs and they wanted Dad to buy them a bomber.

Mr. Kirkwood asked what I thought of their great idea. He found out fast. I told him I had a better idea. The boys should run off from home and be cowboys.

PYLONS

There is an old saying while practicing pylons or turns about a point—do not use a cow. She may walk into a barn and you may spin in.

ENGINE FAILURE
1974

The Beech B-65 was en route Instrument Flight Rules (IFR) from Venice, Florida, to Wheeling, West Virginia, on February 24 when the pilot advised Indianapolis Center he was having engine problems and desired to land at Parkersburg, West Virginia. He was turned over to the Parkersburg tower and advised the tower that the No. 1 engine was feathered and that he was having problems with No. 2. As he came over the field, the pilot reported he was too high to land on runway 10 and would turn for a landing on runway 3. Shortly thereafter the right engine quit and the airplane spun to the ground, killing all five aboard. Investigation into the cause of the engine failure is continuing.

SPORTSMAN NO

All of the state papers carried the following story on February 1, 1974:

Man Accused of Hunting From Copter

A hearing will be held next week before Justice of the Peace E. G. Gainer in Elkins for Mr. Coberly of Elkins, who has been accused of hunting from a helicopter and hunting Canadian geese out of season. A day and time for the hearing has not been set.

The charges against the president of the Elkins helicopter firm, Rotorcraft Inc., were made in warrants served by conservation officers Bill Armstrong and Ken Painter of the Department of Natural Resources.

The incidents allegedly occurred January 17, near Elkins.

In 1972, the same man was fined for hunting deer from a helicopter. I personally wish when this happens that the deer and geese could shoot back.

PLANE CRASH KILLS THREE
November 9, 1974

Two men and a woman were killed near Morgantown Saturday when the private airplane in which they were riding crashed near the Morgantown Municipal Airport, according to state police.

The aircraft was reportedly returning to Morgantown from a round trip flight to Pittsburgh. Witnesses told authorities that the plane had passed over the airport, circled, and apparently was attempting to land when the accident occurred.

Lawmen said the plane narrowly missed striking a trailer court before it flew into a cluster of trees, which sheared off the aircraft's wings. Skies were foggy at the time, police said.

LOW FLIGHT
1974

The pilot of the rented Piper Cherokee 140 took off from Ashland, Kentucky, at 5:00 p.m. on the afternoon of June 8, and according to witnesses proceeded to fly over several communities at low altitudes. The airplane proceeded up the Ohio River and was seen to fly under a railroad bridge. It then turned to follow I-64 eastbound out of Huntington, West Virginia, and struck a cliff adjacent to the highway. The pilot was killed in the crash. Parts of the aircraft's vertical stabilizer were found entangled in transmission lines some twenty-five hundred feet from the crash site.

UNIDENTIFIED FLYING OBJECTS (UFO)

UFO I

My personal belief is that we have had no visitors from outer space except those mentioned in the Bible of spiritual beings. Nevertheless, to make a positive statement this day and age would be foolish.

The things others and myself have pondered and wondered over so long, without a possible clear explanation, leads me to believe in atmospherical phenomenons. A transversion of magnitude, such as the northern lights, the reflection as one mirror to another, stopping at a certain point, creating an illusion.

Why are UFO's seldom reported during daylight hours? This is mostly because of the greater reflection of power of light after the hours of darkness.

With all of my unbelief, here goes my unexplainable sightings:

In 1948, W. R. Wise and myself sighted a bright, fast-moving object some two hundred feet over our heads. This time it was no phenomenon, probably a meteor. Hours later, I discovered, lying scattered across the airport, metal particles resembling lead. It was as light as aluminum. Mr. Wise, being an engineer, rejected any possibility that this came from the UFO or meteor.

I then sent a few pieces to the Smithsonian Institution, Washington, D.C. It was returned and they asked that I send it to a research center in Dayton, Ohio. I did, without response.

UFO II

Other such sightings appeared the night following the launching of the satellite Echo. The visibility was not good. We were watching the skies with anxious eyes, when directly over our heads appeared a cigar-shaped object. The size and length, if it was two hundred feet high, would compare with a telephone pole; but supposing it was thousands of feet high, it would be thousands of feet long. It glowed with the color of a fluorescent light.

UFO III

There was a full moon, and Roy Swanigan and J. Z. Summerfield were about to take advantage of the clear, quiet night and go for a sight-seeing ride in their PA-14 Piper. Appearing on the horizon was a glow which appeared to be a building burning. Summerfield and Swanigan said they would investigate.

The glow, which looked as a building burning, rose in the sky as it traveled from northeast to southwest. It was twelve minutes before it disappeared. Summerfield and Swanigan returned empty handed. They did not catch it. Summerfield, the next day, asked what that was last night. I suggested it may have been that the Good Lord had taken his own and departed. His reply was, "I did not see anyone missing." I told him that he may not where he worked.

UFO IV

In 1967 I was returning from Baker Field, Burlington, West Virginia. The flight was returning after having taken two boys to a camp. Their mother went along for the ride. As soon as we were airborne, I made an attempt to contact Morgantown radio to file a flight plan. I was told to keep the radio silent because Morgantown flight service radio was talking to another aircraft which was over Cumberland, Maryland.

I turned both radios on to obtain the conversation of pilot and flight service. These are the exact words of the pilot, "We are 5,000 feet high. The UFO is pear-shaped. It's height appears to be about 60 feet. Has a railing around it with several port holes above the railing. When I fly towards it, it puffs off some steam from the bottom and it rises up several hundred feet." Flight service then advised the pilot to return to his home airport; that a jet from Pittsburgh would check it out. I very much wanted to reverse my course and go see, but with a paying customer, a lawyer's wife, you don't go chasing UFO's.

UFO V

One other strange sighting occurred late one night. I called my sister, "Look at that. Did you ever see anything to compare." She turned to go back into the house, saying that it was only the moon. "Which one," I asked, "we have two tonight."

This was probably a rare case of ice crystals setting up a mirror reflection.

I have seen the same from the sun. You could not distinguish which was the actual one. In this case the visibility was very poor.

UFO VI—September 13, 1952

A meteor flashed through the skies and was seen by thousands of people simultaneously.

Seven persons, six of them young men and one elderly woman, reported seeing a monster with a red face, green eyes, and a tail like an alligator. This was in Braxton County, near Sutton. One young man was treated for shock and was revived with smelling salts. The young men stated, "We saw the fire fall from the sky and it come down in the hills. We started out to look for it. It was night and as we approached the hill they saw lights flashing on and off. Reaching the summit of the ridge we smelled a horrible odor; it smelled like sulphur." The police were notified and reported the young men were trembling. Several men went back to the scene and reported there was a sickening, unidentifiable odor. There were no signs of the monster.

The next day, September 14, the story broke in all the newspapers. Being interested, I took a carload of explorer scouts to search the area, finding only what appeared to be where a giant tripod had been placed. The holes were thirty feet apart, four inches deep in the ground. I have no further comment on this sighting.

UFO VII

While waiting for the 11:00 p.m. news, I turned to a station near Clarksburg. The music stopped for what they described as a special news event. His words were as follows: "Fifteen miles south-east of Clarksburg, a UFO has landed. Their newsman was on his way to the scene. Stay tuned for details." This report was given several times. Then they reported there were two deceased in the UFO—very small creatures compared with man.

The next report came hours later. The area had been sealed off by Army and National Guard troops from Kingwood. There seemed to be a news blackout. Not one word further came out on this. Whether this was an H. G. Wells type stunt or the real thing, I cannot say. The next day I went to Benedum Airport. They were non-commital, flying the area eight miles southeast of Bridgeport, West Virginia.

Flying low over the departing convoy, I sighted several army trucks on an abandoned farm. One large flatbed trailer truck had a circular disk covered with canvas. The disk appeared to be twenty-five feet in diameter, four feet thick in the center, and this was followed by a truck with a large crane. Twelve trucks, and four jeeps were leaving the scene. Where they were going and what they were doing there, I cannot say.

One week later, I drove to the farm to satisfy my curiosity. I found only one peculiar thing—a three to four foot gash in the clay bank hillside as though a meteor had struck at a forty-five-degree angle. There was no sign of heat or burnt place.

For those who have never seen a UFO, keep looking.

UNIDENTIFIED FLYING OBJECT SIGHTED NEAR BECKLEY AIRPORT
October 16, 1973

Virtually all the employees at the Raleigh County Memorial Airport on Monday night said they spotted an unidentified flying object in the sky above them.

According to Howard Moneypenny, a weather service specialist with the National Oceanic and Atmospheric Administration, the object appeared first at about 8:45 p.m.

Characterized by flashing red, green, and white lights, the object was in the area of the airport for about thirty to forty minutes, Moneypenny said, and at one point it hovered for about fifteen minutes at a point three hundred degrees west-northwest.

"It had no definitive shape," he added, "and I have no idea how far away or how big it was. Our visibility was unlimited at the time and there was just no way of telling."

The airport has no radar unit.

A local pilot, upon sighting the object, took off in a small aircraft in pursuit of it. Although the pilot was able to keep the object in sight for several minutes, "It just kept moving away."

UFO

Walking into my office was a former student. He was well on his way to success in aviation as a co-pilot for a major airline. He asked if he could talk with me in strict privacy, and he asked that I not reveal the source of information of our discussion. He said he was troubled because he had seen the impossible while flying at thirty-five thousand feet.

As he continued his story, he stated the captain and at least six other persons had seen the same. I asked him if he was speaking of a UFO? "Yes and no," he replied, "You will make fun of me but how could six other people see the same thing?" I asked him to continue and try and see if I would believe him. He said he needed to tell someone. This had completely dominated his thoughts since the sighting. I told him I would try to understand. As he continued it was not easy to tell that it was fantastic but true. He continued, "We were level at thirty-five thousand feet. Before our eyes appeared a ship—and I don't mean an airship. The captain, with a sixty degree bank, turned to miss it. Back on course, neither of us spoke for some time. The stewardess came forward asking what the steep bank was all about. She had spilled a pot of coffee, and the passengers were talking silly talk as to what they had seen. The stewardess was asked not to discuss this but to ask for cooperation from the passengers and to collect written statements as to what they saw. Three of the six passengers complied. We now had five unrehearsed and undiscussed statements. The only deviation to the written statements was that each version disagreed as to the size and distance. All other descriptions were amazingly the same, including the captain's and myself." He asked, "Now, what do you think they were? Derelict ships floating up there at 35,000 feet?" My answer was, "No. They would be clearly visible from the ground." I assured him I believed him. I asked him to consider the mirages in the desert and the transversion of a deposit of magnetic air. I told him of the time, January 25, 1971, at night, when I was traveling from North Carolina, crossing the Allegheny Mountains. We were over Princeton, West

Virginia, and Bluefield was clearly to our left where it should be. Looking only past the city limits of Princeton was Lewisburg, with the lights brighter than Princeton. Lewisburg was fifty miles east. Bewildered by the phenomenon, I decided to fly toward it, to see if it moved away. Its position then appeared to be between Princeton and Athens, which could not be. As I approached the light I could have counted every light as the runway of Lewisburg Airport. At the time we were supposed to be over the atmospheric phenomenon. There was a flash of light as though someone had snapped a flash bulb camera in our faces. This light I believed to be the reflection of our own airplane's position lights on our wings against this magnetic clear air of which it had magnified the light several hundred times. The town of Lewisburg had disappeared, and looking in its direction was only a faint glow on the horizon.

I then told him, "Back to the mystery of your ship at 35,000 feet. If you had tried to ram it, the light from your plane would have nearly blinded you, and you would see your ship no more. You would have only passed through air. I could not say whether the ship you saw was in a harbor a few miles away or on the other side of the earth. As you ponder over this transversion of magnetic air phenomenon, consider the northern lights and how they reflect in the upper atmosphere. You actually see a picture of iceburgs."

John Grimmits, a local pilot, told the story of a happening while he was on board a naval vessel. The radarscope picked up a harbor thousands of miles away, in North Africa. This is, of course, most of the UFO's.

PHENOMENON

The last sighting of aurora borealis (northern lights) in this area occurred on February 11, 1958. There is no definite time for the phenomenon to occur but it does at the time of greatest sunspot activity. The displays probably occur when protons and electrons are shot from the sun, striking the earth's upper atmosphere. The earth's magnetic field directs the particles toward the magnetic poles. As the particles move, they collide with atmospheric particles and change their electrical charge and glow.

To make a long story short, we do not know what causes the phenomenon.

UGO

UGO—Unidentified Ground Object. On a return trip from Elkins to Fayetteville, we sighted a UGO lying on the edge of the Monongahela National Forest in a small clearing. We estimated its size to be fifty feet high and twenty feet across the bottom, tapering up like an old-fashioned ice-cream cone, with what appeared to be an eight-sided door on the top.

After returning to report the sighting, Emil Jurak and Herbert Reece departed to have a look. Upon return, they said they believed it to be a hunting camp. In later weeks I have searched for it many times, unable to locate it. Yet I was able to direct Mr. Reece and Mr. Jurak to its location without difficulty.

PERMITS

I notice that they now issue a certificate of authority to take weather observers. Isn't that something? They are licensing the weather guessers.

It looks like the time is near at hand when the bureaucrats will require you to have a permit to use the bathroom.

1961

I had a warm experience while flying a blind lady to Charleston to take an airline. The lady had so much luggage that I could not lead her and carry the luggage. When some of the Kanawha Airport attendants saw my difficulty, they came running, took all the luggage and opened the doors ahead of us. They were a lot of help—those friendly Charlestonians.

1959

Landing at Lunkin Airport in Cincinnati, my passenger—a total invalid, was being transported to a doctor in Cincinnati. I called a taxi. This man was able to set up, but unable to move. I could not find an attendant to open the gate to bring the taxi to the airplane. Opening it myself, immediately as the cab entered, I was told to get it out. My reply was, "I have an invalid in the airplane. Will you help me?" His reply, "Yes Sir, I sure will."

APRIL 4, 1974

After completing the 1944 tornado story, 1974 tornados took 320 lives and hit five states. Only one death was recorded in West Virginia—that of a little girl when it is said their mobile home was blown seventy-five feet into the air. This was at Meadow Bridge in Fayette County. Xenia, Ohio, was fifty percent destroyed with some rubble picked up by the storm and carried two hundred miles.

NOVEMBER 4, 1975

As the Beech King Air C-90, piloted by Steve Poe of Saint Albans, rolled down the runway for takeoff at Kanawha Airport, the engine failed before he was airborne. Witnesses saw gas streaming from one side of the aircraft. The pilot aborted the takeoff. The plane rolled two hundred yards over a steep hillside before coming to rest. Six lost their lives.

THANKSGIVING DAY—1977
Beckley

The Piper Cheyenne, during a dense fog, crashed near the Raleigh County Memorial Airport. Daugherty, FAA agent, said the airplane made a successful approach to the runway. He also stated that pieces of the gear were found on the runway. There were marks where the propeller struck the surface.

The preliminary findings indicated the airplane traveled eight hundred to one thousand feet on the runway and one thousand feet on the sod before going back into the air, crashing one-half mile from the runway. Six died in the crash.

HE SENT ME BACK
1967

Down, down, my airplane Piper PA-22. Out of control, it seems as though we have had it, probably thirty years of flying finished.

We were fifty-five hundred feet in a flat spin caused by the vortex wake turbulance of a B-52 bomber.

My thoughts turned back to the evening before, at close of a late service, with Rev. Richard Miller. We stood in front of a little country church discussing at length God's blessings in my life. I told the Reverend Mr. Miller I had prayed long that if an accident or misfortune fell on our little flying service let it be me. Not my students. Little did I know before the sunset of another day I would search my heart to see if I meant this.

Until this time we had advertised the nation's best safety record. God had certainly stood by our work.

Saying goodnight to the Reverend Mr. Miller, "I have an early cross country down Nashville way."

Long before daybreak we departed Fayette Airport. My passengers were Mr. Gieseking and Mr. Renick. They were students taking advantage of a business trip to log flying hours.

Charleston radioed, "This is 1782P, go ahead 82P." "How's

the weather in Bristol, Knoxville, and Nashville?" The reply, "clear visibility, unlimited." "Roger, that is our route."

The first clear rays of the sun found us over the Blue Ridge Mountains and the beacon yet visible at Tri-City Airport near Bristol.

Much of God's magnificent wonders had passed beneath our wings. Landing ahead of schedule a few hours, we were homeward bound. Our returned trip direct, weather nearly perfect. We were fifty-five hundred feet; a time to relax.

Suddenly without warning the plane went into a flat spin. Descending rapidly, the controls were dead. No effect as if the cables were cut.

DOWN, DOWN, much to be done. If only my passengers would come through this safely. Time to whisper a prayer. "O God spare us, at least spare my passengers." All conventional recovery procedures were tried, but failed. Prepare for crash.

There was a little sign in the panel of the plane. SPINS PROHIBITED. I always thought this meant aircraft was not structurally safe for spins. Little did I know spin recoveries were difficult or impossible. Mr. Renick, beside me, put his feet on the instrument panel. I will, also, when closer to the ground. This should lessen some of the shock of the crash. All switches off, cut gas off. Braced for the impact and surely death, my eyes saw a familiar sight, a little testament I always carried in the plane to read, when added strength was needed. Then the thundering crash.

All was silent except for something that sounded like rain. It was gas coming from the nearly full gas tanks. Renick was struggling to free himself, something was burning his leg. I supposed it was a hot ignition wire, which would soon ignite the gas. If this happened everything would be turned to cinders.

Mr. Gieseking was out by his own strength. Mr. Renick loosened his safety belt, holding his hands high above his head he was pulled to safety (it was only hot oil burning his leg).

Mrs. Rubin Decker, whose yard we had crashed in, was gently pulling me free from the wreckage. Unable to clearly collect my thoughts, elapsing into unconsciousness near the valley of the shadow of death.

Then a dream, or was it a dream? I was traveling toward a beautiful light, with the soft colors of a rainbow. When reaching it I was stopped. A soft voice said, "Step through it, on the other side is eternity." "Or would you rather return and finish what you were born and sent to do."

I asked if there was someone else that will do the work. The answer was no, but it is a long way back. I said send me back.

And then I awoke to reality in the emergency room of the hospital at Somerset, Kentucky, with fourteen bones broken. When they finished in the emergency room, I asked the nurse for a pen and paper. I wrote the following:

> He sent me back, for a time to face man-kind.
>   to do the work and even more that I
> should have done before. With a faint heart,
>   if I could be free, one step forward
> in eternity I would be. It must have been an
>   Angel that said to me "It is a long way
> back, to finish what was meant to be."
>   That is why I am here today, to help some-
> one along his way.

A year later and much to be done. Recently I stood before the golden flame of a candlelight and watched young men of my Scout unit light twelve candles, symbol of the Scout laws. Each saying the Scout Oath, "I will do my best to do my duty to God and my Country."

Now I know in part why he sent me back, to finish what was meant to be.

The steel braces on my legs are no burden. For one touching so near to eternity. With humble gratitude that my passengers were not seriously injured.

Within me a burning desire to help the youth of our country. Teach those to fly who have a deep sincere desire that they may go into all the world.

Mr. Jones, I am back.

JULY 1967

Finding myself confined to a wheelchair or bed after my crash of 1967, I felt compelled to return to the airport. I prevailed on others for transportation. I could not be a burden to those at home. The situation at home was a sister overworked caring for our invalid mother who was slowing passing away. If there was one bit of strength left in me, I could not impose on her. I was near bankruptcy for my business. I badly needed additional medical care. Finances were at an all-time low with me. With determination as I had never had before, I would not spend one cent that was not necessary. If everything else was lost, I would keep the airport property. My purpose at the airport was not only to manage, but for quiet, peace and rest, and also not to be a burden to others.

With fourteen bones broken or fractured, the days were long and painful. After the third day at the airport, I knew I must make a decision—no more pain pills before I became hooked on them. There are no words to explain the sudden withdrawal from those pills. Only those that have experienced such know the feeling.

This very day, the worst of my life—bad luck. This can only be told because there are witnesses. In came the FAA bureaucrats, finding me lying on a couch poking a few crumbs through a broken tooth. My jaws were broken and my teeth wired together. One reached to shake hands with me, and not realizing that he would grip my hand firmly, it rebroke a finger.

The FAA demanded to see some papers. The agents actually picked me up, put me in a wheelchair, lifted me up the steps into the office and demanded to see the airport records. I was questioned and browbeaten for hours. I was nearly at the point of collapse when the hour of 5:00 p.m. came (meaning the day was done). Some old custom of no work after the hour of five handed down from the gods and godesses of bureaucracy.

They gave me orders to be at the airport office at nine the next day. I was depending on others for transportation. My sister, realizing what I had been through, called them and told them that I was not in shape to make the trip. Besides, it was raining, and in my weakening condition we were afraid of pneumonia. She invited them to the house. There, before my sister, they were perfect gentlemen.

My brother, Malcolm, came to my financial rescue, allowing the poor man's flying school to continue.

It is hard to understand, but two county officials went to Washington, D.C., to try to obtain a county airport which would put me out of business. All of this happened while I was in a wheelchair.

FINAL REPORT ON ACCIDENT
June 12, 1967

DEPARTMENT OF TRANSPORTATION
National Transportation Safety Board
Bureau of Aviation Safety
P. O. Box 1245, Miami International Airport Branch
Miami, Florida 33148

January 4, 1968

Mr. Frank Thomas
Box 464
Oak Hill, West Virginia

Re: Accident near Barrier, Kentucky, 6-12-67, Piper PA-22-150, N-1782P

Dear Mr. Thomas

Enclosed are documents held by the Board in the conduct of the above referenced accident investigation. The return of these documents constitutes the return of all material held by the Board relating to your accident.

In the event you have not received a copy of the Release of Aircraft Wreckage from Mr. Edward Boss, Operator of Cumberland Aero Service, Somerset, Kentucky, where your aircraft was stored, I am enclosing a copy of same.

At the present time, I am at a loss to determine the causal area of your accident. Please be assured that I have investigated every possible area.

Since your detailed description of the flight conditions encountered led toward a structural or control surface problem, our efforts were concentrated in that area. However, no malfunction or failure could be found.

The Board's sole purpose is to determine the cause of accidents in order to prevent accidents of this nature in the future. Since any investigation can be reopened should significant and factual evidence become available to determine the cause, we welcome any evidence you may feel would add to the determination of a probable cause.

Sincerely yours,
Robert L. Oelker
Air Safety Investigator

MEANDERIN' AROUND
1946

Jack Johnson was a newspaper man, and he gave us our first public acknowledgment that we were in business. This helped much for a new struggling airport. We were deeply grateful.

Frank Thomas, the young man who owns the Fayette airport near Fayetteville and who can make an airplane "talk," said yes, he'd take me to Richmond. I had been unable to get a train reservation, and having at least thought I was too sick to sit up all night, I was "up a tree" as to how to fill a Monday afternoon appointment at the Medical College there.

Now I'm a complete novice about this flying. I had been up in an airplane, an open cockpit affair, but admittedly had spent most of the time up there wishing they had blindfolded me and wondering how it would feel if I should fall out. It will sound strange to the many young men who now fly or have flown many, many hours when I confess that the experience was not completely pleasant or comfortable.

But Frank talked about a trip to Richmond like I talk about crossing the street to get a cup of coffee, so on Monday morning about 9 o'clock I stored my traveling bag back behind the rear seat and we took off.

The two-passenger Aeronca, in case you hadn't noticed, really looks none too big just sitting there on the ground. And on the inside, there is room for just two people, those two and no more. A window is right up against you on each side; that's how wide the ship is not!

The takeoff was easy. Frank does that with the same nonchalance as you would enjoy in tieing your shoe. He just turns on the juice, the motor gets going like a scared rabbit, and the next thing you know the ground is down there somewhere, and you're up here somewhere else. It is like, under Franks careful guidance, like being lifted from the floor onto a feather bed.

Within seconds I got my first real thrill. Now we were flying over the treetops east of Route 19-21, and then, all at once we were over New River gorge right above Brooklyn! Hundreds of you people have taken that trip, of course, but for those of you who have not, let me go on record right now as telling you that you've missed something and that you'd better get out to that airport and make up for it. For there's no spot, I can tell you, between here and Richmond that's as beautifully rugged as that massive canyon seen from the air. It actually takes your breath away.

We headed for White Sulphur. Frank flew by mountain tops rather than bends in the road as I was accustomed to traveling. There, he pointed away into the distance, is the peak where we enter Greenbrier county, and we did, by golly, right there, half an hour later.

That's Alderson on your right, he said, and within another few minutes pointed out Rainelle on the left. That, I am forced to conclude, is taking a long look. Lewisburg's fairground was beneath us within 35 minutes, and that despite a strong headwind, truthfully, I would have known nothing about the headwind had Frank not told me, but I feel like an old navigator talking about it now.

There was considerable fog, the weather was none too good for flying and there were some bumps. All this, of course, I also know because Frank told me. As for me personally, everything was just as it should be. I suspect I would have thought it a part of the regular run if we'd landed in a treetop!

The James river was our guide from White Sulphur on, and it was there that I was reminded of how kids take short cuts, even across city lots. For while we followed the river, actually we didn't follow it. Frank would look away out there at a sweeping bend, then he'd just cut across the hill. That, I surmised all by myself, is really straightening out the river.

We stopped at Lynchburg for fuel, setting down at the beautiful port with as much ease as an eagle, and then skimmed, with better wind but with a low ceiling, into Richmond in about an hour more. It was really slick going down that valley. Richmond was actually beneath us in less than three hours of flying time, and I was filling my appointment within four hours after I left Fayetteville.

That is not, of course, fast flying in these days of bullet speed, but it's fast in my book, and I heartily recommend it as a way to get where you're going without being worn out when you get there.

Oh yes, if you've never flown before, I warn you that that Aeronca will seem frightfully little when you're up there above those mountains, and you'll look in every direction for a place on which you might land in case something should go wrong. Actually, the higher the mountain the smaller that ship got for this nervous customer! But now I want to try it again, and will, so help me, just as soon as the opportunity offers.

And I can also tell you, in case you're thinking of trying that New River gorge trip, Frank's infectious grin and his complete confidence in handling that ship will very quickly take care of any nervousness you may have. So go take your look; you'll remember it as long as you live.

I admit that lots of people have moved to the cities of this vast land, and on occasion I have been forced to think that literally everybody is right there on Main street every Saturday night. But my little trip over the mountains to Virginia last week convinced me that there are still some folks living in the country, and I mean country.

For believe it or not, there are many, many houses away back there in those mountains, houses which to all appearances are completely isolated, and I saw no less than half a dozen from which nothing more than a path seemed to lead. If there was a road within ten miles of them, I just couldn't find it, and I was looking for roads.

Like almost everybody who got himself caught in the maelstrom of industrial living, I have often had a yearning to get back there in the wilds where none was but the old man himself, there to forget there was ever a radio or newspaper, there to go possum huntin' when I felt like it, there to disremember to shave, there to sleep until the mighty sun climbed high enough each day to get over that big mountain to my east.

Of course, I've known all along that it wouldn't all be quite as nice as I pictured it, but I had to take that trip to know very definitely that such living is not entirely what it's cracked up to be. Yes, there is quiet, and I could very well understand that one would not be disturbed for days on end except by guys like Frank Thomas and me coming over in that plane, but I could also see from the air that there were other things also. And I've never had those things so indelibly impressed upon me.

In the first place, it was much more than evident that there were no bathrooms. Proof of that was in the fact that there was a path leading from each house; further proof was that in most instances if there was water even close, I couldn't see it. In only one instance was there a stream, and it a small one, within a hundred yards of the home. I presumed, of course, that springs furnished the water, and I further presumed that they meant a water-carrying job which just about no man likes.

At one spot it was washday. I was surprised that that was not universal, since we made the trip on a Monday, but it was not. Else we got out too early for the clothes to be on the line. But at this one spot some of the clothes had been hung out. But the remainder of them were being washed, so help me, down there in that little stream, either on a board or a flat rock. I could not tell which from a thousand feet up. But there, in as primitive a manner as any of your ancestors did their laundry, some woman was battering out the dirt.

Now that was not too bad, I well know, on Monday, September 23, when the weather was balmy, but believe me, no little amount of it would be at all pleasant a month hence, and the carrying of the water to be heated in that kitchen would be something less than fun, especially when the weather acts up.

The one thing I did observe which looked just about as all right as it could, was that with few exceptions the men of the households appeared to have nothing to do except just sit there. I do imagine they whittled.

But I do swear that in the four of five instances when I could see menfolk about, not a single one of them was ambitious enough to wave at us, although the womenfolk did lift a handkerchief as we passed over. I was reminded of the Collier's cartoon where the hillbilly and his dad lay beside the creek with their feet in the water.

"The crick's a risin', pa," the son was saying, "it looks like we're goners!"

I can well believe that women do most of the work around

most of those spots, too, because there just wasn't any evidence of man's labors in the way of tilled soil. In two or three instances, there wasn't enough cleared land to raise a disturbance on, possibly because the old man and the sons didn't want any of the possum timber cut!

But whether there was cleared land or no, the meals had to be cooked, though heaven knows where they came from, the washing had to be done, the fires had to be built. And I know full well the snow gets deep on those mountain tops long before Christmas.

I came back somewhat cured of my yen to "get away from it all." I don't like the cities, that I insist upon, but there's a happy medium, believe me. And I'll stick with it. There could be worse things than being back out there atop a high mountain with nothing in ten miles except air and raccoons, but after that very complete look, I can't imagine what it would be.

THE LAST OF THE LITTLE STRIPS
Sunday Gazette Mail **State Magazine, December 3, 1967**

by William C. Blizzard

In nature, the hardest boulder stands longest against the forces of erosion, and may one day be seen as a solitary natural tower on a hilltop, silhouetted against the sky.

It is the same with men and the institutions created by man. Forces of social erosion are constantly at work, and only those men and institutions survive that are tough enough to withstand such social erosion.

The hardest granite is not left unscarred by this process of attrition, nor is the most flint-like man. But the mere fact that they survive as long as they do is wonder enough.

Some such thinking probably was in the mind of the philosopher and Latin scholar who penned the immortal slogan, "Illegitimati Non Carborundum." Translated and bowdlerized, this reads, as you may know, "Don't let the illegitimate ones grind you down."

Frank K. Thomas of Fayetteville must be a hard man who

doesn't grind down easily, for he has survived twenty-two years as the owner of a small, private airfield, the lone small-business boulder of this nature that is silhouetted against the West Virginia sky.

"When I started this Fayette airport," he told me, "there were 57 small airports in West Virginia, including this one, and most of them had never received a penny of state or federal funds. As far as I know, I'm the only one left today.

"By that, I don't mean there aren't some other small, privately owned fields left in West Virginia. There are a few, and they're doing good jobs, like the one at Pence Springs, or Stewart Airport at Parkersburg. But all that I know of have received some sort of government funds. I haven't."

Thomas didn't specify what he has against governmental money being invested in his airport, at a time when it is common knowledge that all major airlines are heavily subsidized by federal funds—that is, by you and me. It is apparently a matter of philosophical principle with the Fayetteville businessman.

Frank Thomas is as firm a believer in that old-time, small-business, free enterprise unsullied by governmental interference as a Biblical fundamentalist is a firm believer in that old-time religion. Though the stormy crasher of giant mergers are heard in the land, and governmental ogres batter him, he stands unmoved, a free-enterprise, rugged-individualist who asks only for laissez-faire.

Needless to say, his concept fits perfectly the image of capitalism in a nineteenth-century, New England small town, but is fast becoming an anachronism in modern America. As is obvious, small business is being gobbled up by large in every field of endeavor, and even the largest businesses are merging to become, in effect, one.

It is tempting to speculate as to the nature of the offspring of the union of two giants such as Continental Oil and Consolidation Coal, for instance, but the risk of sounding like Eric Sevareid in his more pompous moments is too great.

To keep from being exterminated entirely, small businesses have been forced to seek governmental aid, and federal money is poured into major airlines as an alternative to what is seen by most lawmakers as the greater evil of out-and-out government ownership.

Such is the tide of the times in modern America, and such is the tide Frank Thomas of Fayetteville is swimming against—as he well knows. The struggle, he admits, has made him bitter.

He will tell you quite frankly of his battles with the Fayette County court.

"I've heard stories," he said, "that they're thinking of condemning this airport. We don't have a county airport in Fayette yet, you know, and that would eliminate competition. But I'll do all right if the county court keeps its red, socialistic hands out of our business."

Needless to say, the Fayette County court would be as much surprised to learn that it is socialistic as would be the various socialist countries in existence today. As Thomas says, his battle to keep his airport alive over more than two decades has made him a little bitter.

Thomas not only does not believe in governmental subsidies, he is not one to borrow money except as a desperate last resort. He is, apparently, a pay-as-you-go man. His reasoning is as follows:

"I could," he said, "borrow money and build one of the world's finest small airports here, but the kind of taxes that would be imposed on me would be totally and completely out of reach. As it is, my taxes here have been increased 16 times on part of my property and 24 times on another part."

Thomas, who is now forty-six, was taught to fly by the late Herb Sessler of Mount Hope, who first operated an airport near the Raleigh-Fayette border in the thirties, then moved to the Beckley-Mount Hope airport about 1937.

Sessler later began an aircraft factory in Florida, then started a similar business in Princeton, West Virginia. He died in Millburn, Florida, in 1952.

Thomas received his commercial license under Sessler in 1942 and his instructor's license the following year. He was not in the armed forces in World War II, but did some pilot training in Danville and Roanoke, Virginia. He later flew for Sessler in Beckley, doing charter service (in effect, an air taxi) during the war years.

"My longing," he said, "was to establish an airport of my own. I began here in the last days of 1944 and in early 1945."

True to his do-it-yourself spirit, Thomas began his strip by cutting down trees with his own hands, before the advent of the portable chain saw.

"But," he said. "I always took about four axes into the woods with me. People around here were interested, and friends would come to visit. We would talk and chop, and this made the work more interesting."

He took a job as an ore mixer at a metals plant at Alloy, working on his airport dream at the same time. The educational section of the so-called GI Bill of Rights then became operational for veterans, and gave the newborn Thomas Flying Service, otherwise known as Fayette Airport, a start.

"This GI program," he said, "was extremely valuable for people wanting to learn to fly. Many people did learn to fly under this program that work as pilots for major airlines today. Of course, we needed everything. We needed hangars, we needed rest rooms, we needed airplanes, and we needed offices."

Thomas appears to have borrowed money, despite his scruples, to get started. 'Getting what we needed," he said, "put us badly in debt. We didn't always eat, but we enjoyed our work immensely."

There is an airplane hangar on the field and several other buildings.

"I hauled in the rocks and other building materials and did most of the carpentry and other work myself," Thomas said. "I also let students trade their labor for flying time."

He still does this. To augment his income he began an ornamental-concrete business on the field seven years ago. Young students work at this and Thomas pays them in flying time.

"It's dirty, hard work," Thomas said, "but nothing worth while comes too easy."

For recreation, Thomas, who is a bachelor living in quarters at his airport, headed a local Boy Scout unit for seventeen years. On June 12 of this year, however, bad luck ended this activity and very nearly ended Frank Thomas.

"Unquestionably," he said, "our airport had the nation's best safety record until June 12. But on that day I was flying two student passengers near Monticello, Kentucky, and we struck the vortex, or wake turbulence, of a jet or jets. My students were badly bruised, but I got two broken legs out of it."

Because of his broken legs, Thomas resigned his scout work. "I don't imagine," he said, "that I'll be able to do as much hiking as I used to."

On the day I visited him, however, he was quite happily flying, giving instructions to students despite casts on both legs that forced him to use a light-metal walker to hobble about the field.

"People ought to know about these jets," he said. "A light plane can hit their wake turbulence 20 minutes after they pass over, and it's like hitting a stone wall. It caused our first accident in 22 years. There wasn't 15 cents worth left of our airplane."

At present, Thomas owns four airplanes, of the "Cub" variety. In his twenty-two years of flight instruction at Fayette Airport, just under six hundred students have been soloed successfully.

He does not have a charter service now, and airport income is derived from student fees, gasoline sales to owners of eleven private craft stationed there, flight fees for pleasure hops over the New River Gorge, and small profits from the ornamental-concrete business.

What might it cost to learn to fly at Fayette Airport?

"I don't want to tell you what I charge just now," Thomas said. "Just say this is a poor man's flying school. I don't want a big rush of business until after December 1. We aren't closed many days of the year, and after the first of December I should be in good shape and ready to take on all comers. Last winter, we were actually operational more days than either Beckley or Kanawha Airports."

Thomas's plans for the future include the purchase of two more airplanes, the building of another, east-west strip, and ultimately, the paving of his present north-south runway.

There is, as is necessary in a man able to withstand the erosion of time and the buffets of adversity, something of the zealot in Frank K. Thomas, and he doesn't deny it.

"I've told people many times," he said, "that if ever a preacher had been called to preach, I've been called to fly."

FORCED LANDINGS

At the time of this accident, my broken legs were still in casts, due to an accident in a Tri-Pacer. I was riding in a Piper Cub at Fayette Airport, taking off south over the high trees, when the engine failed causing a complete loss of power. The rule book says in the event of power failure to land straight ahead, but there was no place to land—only tall trees. I told Mr. Weeks that we would be just as well off to try a turn. If our altitude allowed us to complete the turn it would be a fast downwind landing pushing us all the way. The turn was accomplished. Our airspeed looked good. The ground speed was very fast and we were losing altitude fast. The treetops were brushing our wheels. We were low and fast and the runway was under us. With this much wind, we wondered if we would stop rolling before we reached the fence. Almost unbelievably, we stopped rolling near where the airplane was to be tied down.

The cowling was removed and we found that one cylinder had cracked and broken loose from the engine case. This was the third forced landing I had had since the crippling accident.

This called for some soul-searching. There had been some silent learning. Were these forced landings warnings, or old age of the airplanes? My decision—the old planes must be sold. They must go, but not me. My enthusiasm was grown. The call to fly was there.

The Battle
with the Bureaucrats

1952

Our visit from the federal inspectors was in progress. The inspector, Mr. Carpenter, with his notebook and tags, found some petty annoyances with each of our airplanes. According to him, these made the plane unairworthy and grounded them.

Harold Stewart, our full-time mechanic based on Fayette Airport, called me aside and told me that Mr. Carpenter's airplane, a Stinson, had a mandatory bulletin which had not been complied with. I, as airport manager, grounded his airplane.

The bus and train service was slow in this area, so Mr. Carpenter reconsidered. Some of the small blemishes on our aircrafts were not as bad as he first thought. He then withdrew the grounding of our Cubs. We also forgot we had seen his Stinson.

MORE THUNDER

T. Thunder Johnson was an aviation agent. One Sunday morning Johnson and Agent Hall arrived at 10:00 a.m. on a surprise attack. As we were preparing for church, Agent Hall, without introducing himself, asked to see my license. First I said, "May I see your credentials of authority?" His reply, "I do not have to show them to you." My reply, "You will if you make an inspection here." He said, "I will get a federal marshal." Again I replied, "The federal marshal will show me his credentials or I won't let him in either." Hall and T. Thunder Johnson went into a huddle. Hall returned with his credentials in his hand, stating, "I suppose it won't hurt anything to show my credentials." I said, "Neither the same with me." He replied, "Thomas, you knew who I was." I answered, "And you knew me also."

They went on with the inspection as time was growing close for church.

Johnson said, "You are doing so much howling about church. I'm a Deacon in my church and I missed church to come here."

I replied, "Praise the Lord. Are you a christian?" He made a quick departure without answer. We were off for church for some badly needed spiritual renewal.

HOW THE INSPECTORS REACT

I. Upon one occasion the inspectors found we did not have our parachute packed in the last sixty days as the law required. He let us know that he would forget this if we would buy an old pistol from him. The pistol was a piece of junk which appeared would kill from both ends if shot. This is the first time I have known of a X ! X ! taking the cash and leaving the gun.

II. In 1958 one of the neighbor's chickens miscalculated its true course and became lost and was found trespassing on Fayette Airport. Without consulting the Civil Air Regulations, we seized the chicken, put it in protective custody and cooked it. This was in 1958, and like 1974, the Civil Air Regulations were changing faster than we would be able to change our underwear. The chicken was quickly prepared for our seldom airport feast. Then the worst of bad luck—in walked Agent Johnson. Johnson, invited to dine with us, was as hungry as a preacher in the middle of the week. He also disposed of one-half of that chicken. Our dinner was gone, but not Johnson. He remained until he found some little gripe upon which he could file a violation. Then he left us with a well-fed, contented look. With his belly full of our stolen chicken, he departed happily.

III. When my eyes behold some of the works of art in the airport terminals, they make artist Moss's West Virginia moon, which he made from an outhouse door, look like a Rembrandt. Also, what of the red marble slabs between the commodes in the little boy's room. There is where your tax money goes.

FROM THE RECORDS

This is what I have been trying to tell you for years—the bureaucracy destroying the small operator.
The record which follows is from Jats West Virginia State Aeronautics Commission, 1952.

AIRPORTS NOT LICENSED

Checking the records and files of the office, it is noted that the following airports have failed to renew their annual license to operated as commercial airports:

Philippi Airport—Philippi, W. Va.
Boone Field—Ronceverte, W. Va.
Black's Field—Rupert, W. Va.
Weirton Airport—Weirton, W. Va.
Fairmont Municipal Airport—Fairmont. W. Va.
Marlinton Airport—Marlinton, W. Va.
McDonald Field—Taplin, W. Va.

In addition, quite a number of operators have quit business in the past year. The following are not operating:

Baker Air Park—Burlington
Clarksburg Aviation Co.—Clarksburg
Martinsburg Flying Service—Martinsburg
Greenbrier Air Service—Ronceverte
Simpson Flying Service—Philippi
Pioneer Flying Service—Morgantown
Air Taxi, Inc.—Parkersburg
Logan Aircraft Service—Logan
Buckhannon Flying Service—Buckhannon
W. Va. Air Service—Clarksburg
Mountaineer Flying Service—Fairmont
O'Neill's Flying Service—Parkersburg
Wings Service—Parkersburg
Valley Aviation, Inc.—Wheeling

Dave Baker at Baker Air Park informs us, "I have sold my planes and am going back to the horse and buggy. I believe this aviation is a passing fancy and will never replace the bobtail horse and the rubber tire buggy."

Surely it's not as bad as this!

MONTHLY INSPECTION

We were proud and anxious always for the appearance of J. A. McCauslang, FAA agent. He was firm and fair and no airport operator could ask for more. On this visit with him, was Carl Peters, the FAA maintenance man. We always knew when they were coming, and we made special efforts to clean the entire airport and airplanes.

In washing Piper 5561H, we found a small crack in the landing gear fitting. We tagged it as, "Not to be flown until further checked." Upon McCauslang and Peters's arrival I immediately showed them the crack. They agreed it should be fixed by a professional. Huffman was the best, his shop being located in Montvale, Virginia.

McCauslang asked to flight check me. From him I always learned much. He had gained such respect that the pilots spoke of him as the "Great White Father." Before flying, I telephoned Jack Wendel asking him to fly the plane to Virginia with someone to follow.

While I was up with McCauslang, Jack came and was propped out, meaning started by hand, by Agent Peters. Upon landing at Montvale, he was met by Agent John Gibson who gave the airplane a ramp check. As Jack tried to explain why he had brought the airplane to Montvale, Gibson, from another district office, proceeded to cause much confusion. I received several letters which I ignored until they were ready for action. When he learned the truth that another agent had started the airplane, how glad he was to dismiss the violation.

PA-18 RADIO
Swallow a Camel and Gag on a Gnat
1962

Such was our opinion of the FAA when one of our PA-18 Cubs was ramp checked at Charleston, West Virginia. The only thing they could find wrong was if you put the stick all the way forward your knuckles would touch the knobs on the radio.

These radios were factory installation. The airplane was grounded and a mechanic would have to sign the ferry permit before the radio could be removed for repairs. Then after the radio was removed, it would be necessary to compute a new weight and balance for the airplane.

After much loss of hours, the FAA decided we could remove the rubber tips off of the radio knobs, these tips being the size of the button on your shirt. Then we would be OK to fly. In evaluation of this, I have never known of a maneuver where the stick would be put all the way forward.

NEW AVIATION GROUP
Aerial Signs Urged

Taken from *Sunday Gazette Mail,* September 29, 1958:

Two letters sent to Gov. Underwood—from Oak Hill and Washington—ask that he dismiss the State Aeronautics Commission and publicly acknowledge the value of aerial markers.

The writers are Frank K. Thomas, operator of the Thomas Flying Service at Oak Hill, and Charles A. Parker, executive director of the National Aviation Trades Assn. at Washington.

Thomas' letter dated Sept. 22, carries the sharpest words. He says:

West Virginia is at the bottom of the list in aviation in the United States. There is one big reason why—there has been no thought given to the choice of suitable members of the State Aeronautics Commission.

"We do not have a single active commercial pilot or airport operator on this board, not even a highly active pilot."

Thomas cites this law:

"... members of the commission ... shall be selected with due regard to their fitness by reason of their aeronautical knowledge and practical experience in the field of aeronautics. In making such appointments, the governor shall, so far as may be possible and practicable, select the several members from different geographical sections of the state."

Thomas continues in the letter:

"Please take note to the fact that two board members are from Wheeling and one from Huntington, which you will agree is not according to the quoted law. This gives the Ohio River Valley good representation, but what about the other parts of the state?"

The Oak Hill operator says it is his plea that the Governor "Take hold of the situation, live up to your oath of office, dismiss the present board and appoint dedicated airmen to this board."

(Members are Hulett C. Smith of Beckley, chairman, and A. W. Paull Jr. of Wheeling, Democrats; Edward F. McKee of Wheeling, A. F. Marshall Jr. of Huntington and State Board Commissioner Patrick C. Graney.)

(Col. Hubert H. Stark recently was dismissed as executive director because of technical mismanagement of aerial marker funds. George W. Hart of Huntington was hired to succeed him.)

Parker of the Trades Assn. tells Underwood in a letter dated Sept. 24, that news of Starks dismissal was "indeed disappointing as we were under the impression that Col. Stark had worked diligently to promote and aid aviation in your state." He adds:

"We are also greatly concerned over the implication that air marking is an unworthy project on which to spend public monies."

Parker says aerial markers still are one of the fundamental aids to safe flying, "despite all the fancy communications and radio navigational systems that have been developed."

He terms West Virginia "a state that needs air marking substantially to improve aviation operations and safety."

The national official makes this suggestion to Underwood:

"We trust that there may be some acknowledgment by your office publicly of the fundamental value of air marking and remove the unfortunate stigma that has been placed upon it in press releases in connection with Col. Stark's removal."

CONFLICT
1969

If you have ever wondered why owning an airplane and basing it on a county or city owned airport costs so dearly, here is your answer in full.

We will look at Kanawha Airport as a typical example. The contract in the past, and as far as we know, now is the same. The contract calls for general aviation to pay a base rent of $18,000 a year plus 25 percent of the gross over $75,000, 30 percent of the gross over $100,000, and 35 percent of the gross over $150,000. Your tax money built this airport. The rich get richer and the poor get poorer.

A part owner of the lucrative franchise, G. W. Jones, Jr., is one of Charleston's two representatives of the Central Airport Authority. The other city representative, R. Q. Jones. G. W. Jones is vice-president of the Amherst Coal Company.

G. W. Jones, being on the Airport authority, and stockholder in Continental Aviation, (dba Kanawha Airport, the same as General Aviation), was listed with two of his cousins as part owner with others in Continental Aviation. The firm has a wholly owned subsidiary, Boone Aviation, Inc., which has entered into a twenty-year agreement with the greater Cincinnati Airport to serve as the fix base operator.

CONTROVERSY

Some local politicians have created much controversy as to the status of Fayette Airport. The local politicians have confused the issue. The following letter will clear the situation up:

Mr. Frank K. Thomas
Thomas Flying Service, Inc.
Box 464
Oak Hill, West Virginia
Dear Mr. Thomas:
 I am enclosing a letter which came today from the Federal Aviation Agency. The letter supplies information which I

believe is responsive to your letter of June 14. You will recall that I replied to your letter on June 18.

The Federal Aviation Agency correspondence indicates that the Fayette Airport, being privately owned, is not eligible for grants-in-aid under the terms of the Federal-aid Airport Program. The letter does say, however, that, under the National Airport Plan, some privately owned facilities have been included, based on their need in the system considered necessary by the Administrator to adequately anticipate and meet the needs of civil aeronautics. The FAA correspondence further states that the Fayette Airport is in this category, but that such inclusion does not represent a commitment by the Federal Government to participate financially in the development of the airport.

You may secure additional information by applying to the field office, the address of which is given in the enclosed letter.

I hope that I have been helpful to you, Mr. Thomas. I assure you that it is my desire to be of service whenever possible.

Sincerely yours,
Robert C. Byrd, U.S.S.

STATE AERONAUTICS GROUP BLASTED BY F. K. THOMAS
1955

West Virginia Aeronautics Commission
Kanawha County Airport
Charleston, West Virginia
Gentlemen:

I fear that there will be an injustice imposed upon the people of the state if I understand correctly your budget for the coming year. Is it true that you seek a $20,000 airplane for the Aeronautics Commission? Why don't you tell the people of the state that the Aeronautics Commission had a twin engine Aero Commander airplane in its name until October of this year? If you buy a $20,000 airplane, why don't you ask for a Rolls Royce automobile to go with it. That makes as much sense.

I am told by your director that there is a member who charges as high as $40 for expenses to attend a single meeting. Why?

With a lack of money to run the State Children's Home at St. Mary's, you should ask for so much?

Let me refresh your memory that 11 years ago there were 52 fixed base operators in our state and now there are only 17 (reference W. Va., Blue Book).

Yours for a fair deal for our state,
Frank K. Thomas
Thomas Flying Service
Fayette Airport
Oak Hill.

WHY NO COUNTY AIRPORT FOR MONTGOMERY AREA?
Montgomery vs. Fayetteville

This was taken from the *Tech Collegian,* an official student publication of West Virginia Institute of Technology, February 25, 1966:

Does Montgomery need an airport? If so, why does it need one? The answer is, it would serve to transport state officials and others who wished to pursue their businesses without suffering the rigors of our state roads. In time of a national or state emergency, it could receive or dispatch aid and necessary supplies. And, with our devious roads, it could save untold time in traveling between two places which might be only minutes apart by air.

The next question in this discussion is: Is it more advisable to build a facility where it is most needed or more convenient? This question has both point and reason in the present state of affairs. It has at last been proposed that Fayette County build an airport. This, in itself, is altogether admirable, but it has also been proposed that the county take over Fayette Airport in Fayetteville as its location. This is less admirable, particularly in the mind of Mr. Frank Thomas, the owner of the airport.

When Thomas started the airport, twenty-one years ago, the runway was 1100 feet long. At present, it is 2200 feet long, with hangar and fueling facilities. These improvements, together with the five airplanes owned by the airport, have been privately financed by Thomas at no cost to the public.

This is the private business that the county proposes to take over. In an interview with Thomas, he pointed out that the takeover would be both costly and senseless. This area is already served by the nearby Beckley airport, and the private Fayette Airport. The local public airport is already losing money, and an attempt to divide the traffic between two airports in such close proximity would only result in a loss to both.

Where, then, should a county airport be located? In Mr. Thomas's opinion, and that of other pilots, the logical location for an airport would be Montgomery. It is a growing city, has a college, and is located near the local factories. Also, as a point of interest, it is the most densely populated area without an airport.

Why, then, is Montgomery not a candidate for the airport? Thomas lays the blame at the feet of the county representatives. In a recent interview, he assured us that the officials were dragging their feet, and would rather put a private individual out of business than work on creating a new facility where it was really needed. In regard to the desires of the people, he told us that he did not know of any private pilots who wished to have Fayette Airport taken over by the county. In contrast, most of them desired to have an airport in the Montgomery area. This would provide a more even distribution of airports and would provide a safe landing place in a busy area.

When asked about the people supporting the take-over of his airport, Thomas informed us that it was backed by the same people who had pushed the defunct New Kanawha Development Corp. In particular, he cited Pat Hamilton, the previous Republican candidate for Prosecuting Attorney, and present Democrat county chairman.

When asked about the management of the airport, Thomas said that he would not serve as manager, as the airport would lose money if located in the Fayetteville area, and the first manager would be fired very shortly after the take-over.

In relation to the Montgomery site, Thomas assured us that he would co-operate in every way possible to make it a success. He said that he would see that it had a qualified instructor, if he had to serve himself.

END OF OAK WILT

This story deals with bureaucrats; facing untrained, uninformed employees of the state.

Such was the morning Mr. Jones, of the West Virginia Agriculture Department, entered my office and asked me if I was going to fly Oak Wilt this year. My reply was, "Yes, if my contract was accepted." He then stated that he would have to give me a flight check. I replied, "I have flown this program twelve years with a perfect safety record. I was not aware that you were an experienced pilot." He then said that he was not, but that he had had two lessons at Huntington and knew how it should be done.

That was the end of my Oak Wilt.

TAX

The following is a documented court record in which the assessor and county court attempted to assassinate my business. Except for the judge giving a fair decision, this would have been the end of the poor man's flying school.

                       Re: Frank Thomas
                       Vs: Appeal
                       Orval Kessler, et als., etc.

Plainly, as will be hereinafter set out, the taxpayer made out a case which required the Board to afford him relief, but none was afforded. The said Board by letter dated April 12, 1965, assured the taxpayer that "after mature consideration of all factors involved", he was entitled to no relief.

Lets take a look at "all factors involved" which were given "mature consideration" by the Board.

Under the heading of "Education and Educational Institutions", Acts of the Legislature, Regular Session, 1963, such act also requires the Circuit Court to act in administrative capacity to make such an assessment as it believes is indicated upon the record made before the Board of Review and Equalization.

In submitting this case to this Court upon the record referred to above, neither side submitted a brief, and argument was informal and very brief. The Taxpayer's attorneys did not stress the "note" referred to in, and attached to, the property record cards. The Prosecuting Attorney simply stated that the Tax Commissioner did not desire to be present nor to present a brief, and that the Cleminshaw appraiser's note had been read and discussed by the Board of Review and Equalization with the Assessor openly and was considered by them. The taxpayer was before them under oath, but no one ever asked him if he had ordered the Cleminshaw appraiser off his premises. In fact he was asked "no questions" by the Prosecuting Attorney (See record page 15), who was present and could have cross examined him. The Prosecutor in submitting the case to this Court said the County Court and the Assessor felt the taxpayer should be denied all remedy provided by law for the correction of any assessment made by the Assessor because of his treatment of the Cleminshaw appraiser in ordering him off his property.

The Commissioners and the Assessor were probably attempting to interpret the provisions of Code 11-3-10 relating to certain penalties and forfeitures and also providing that for refusal to answer or for answering falsely any question asked by the Assessor, or for failing or refusing to deliver any statement required by law the taxpayer shall be denied all remedy provided by law for the correction of any assessment made by the Assessor, etc.

Aside from the question as to whether the Board could properly have considered the note as being any evidence whatever of the taxpayer's refusal, if he did refuse, to answer any question asked by the Assessor, there simply is no evidence whatever in this record that the taxpayer answered "falsely any question by the Assessor, or that he failed or refused to deliver any statement required by law."

The Cleminshaw appraiser was not called as a witness, so counsel for the taxpayer had no opportunity to cross-examine him and thus test the unsworn information which he gave in his note and which was apparently given such great weight by the Assessor and the Commissioners sitting as a Board. This "Note" is the most rank form of hearsay, and apparently, so far as the record shows, the taxpayer was convicted by it, and, even at this very minute, may not know of its existence.

One of the primary causes of the American Revolution, the war in which we won our independence from Great Britain, revolved about our assertion of our rights to be represented in the detemination of what tax and how much tax should be levied against us. I can think of no reason for guarding these rights with less zeal in this day and age. I can think of no reason to harbor less resentment for encroachment upon such now established rights. We are a government of laws, not a government of men. Not only should Mr. Thomas live within the law, but so should the unnamed Cleminshaw appraiser, the Assessor of Fayette County, and the members of this Board. Everyone of these elected officials are under oath to support the Constitution of this State and its laws, *so help them God.*

The law and such Constitution does not authorize them to allow a confiscatory tax burden to be imposed upon any taxpayer because he was rude, or ordered a Cleminshaw appraiser off his property, if he did any of these things. If they want such laws let them take it to the lawmakers.

The firm hired to do the work for Fayette County, is referred to in the record of this case by the Prosecuting Attorney as the "Cleminshaw Company."

The Prosecuting Attorney, at the hearing before the Board introduced into evidence certain "Property Record Cards," which were made up by an appraiser employed by the said Cleminshaw firm; the appraiser's name being nowhere stated in the record. These cards (2 being made up for the Thomas property) were, according to a stamp placed on the top card, "reviewed" "3-24-65" by the "State Tax Commissioner of West Virginia". They were "furnished by the State Tax Commissioner's office" to the assessor's office for consideration by him in making assessments.

These two cards are stapled together and also stapled to them are two small lined pages containing a note pencilled on both sides of the pages. These small lined pages are approximately 3 inches wide and 5 inches long. On the back of the top card reference is made to the "note" contained on the two small pages, as follows:

"Ordered to get
off property
See note attached
to FRT of Card."

The "note" attached is as follows:
"F K Thomas D426 Parc. 03
11.97 AC."
"If the owner of this property should come in on BD. of review, he should receive no consideration whatsoever. That is exactly what he showed me, when I attempted to appraise his property. Before ordering me to "get the hell off my property," he made some very snide and uncomplimentary remarks about the assessor and the sheriff's office. He said that we should get rid of that assessor, because he is favoring the big corporations and picking on the "little guy." He also said that Union Carbide pays the sheriff $1000 a year to get him to keep their taxes down. He also stated that the men working for our company don't know anything about construction and are not qualified to come in here and appraise this property. He asked me if I thought that I was qualified to come in here and tell him what his property is worth. I said "I *know* that I am qualified." Before I could finish, he jumped up, jerked the door open and said "Get the hell out of here."

Because of this, I had to estimate listings and measurements of all buildings on his land. I recommend that they stay that way no matter how much complaining he does.

Incidentally the language that I am using is quite mild compared to that used by Mr. Thomas during our conversation. In addition to following Thomas jumped another appraiser in a barber shop, and definitely stated *bodily* harm to any of our personnel that was caught on any property that belonged to him, regardless whether rental, or owner occupied. When told that his property would be estimated he stated that is allright with him.

The said Cleminshaw appraiser then placed a valuation 28.9 times the value of the 1964 or 24.2 times the valuation for the previous year.

The land books introduced before the Board, but not produced, reflect that this is what happened in almost every case. The Assessor accepted property measurements exactly as reported by Cleminshaw and cut the Cleminshaw valuations in half. The Assessor adopted 1/2 of the Cleminshaw valuation as being 100% of the true and actual value.

The depositions of F. K. Thomas and Aubrey Halstead, a local real estate appraiser, were taken on behalf of Mr. Thomas before the Board. On behalf of Fayette County, the Assessor's deposition was taken before said Board.

Following the Board's ruling on April 12, 1965, the taxpayer took an appeal to the Circuit Court under the provisions of Code, 11-3-25, as amended. This section empowers the Circuit Court to review in a judicial capacity the order of the County Court entered by it as a Board of Review and Equalization.

And that's exactly why the taxpayer has this case before the Commissioners, and he had the right to expect each one of them to live up to his oath, his sworn duty.

What is the value of this land and improvements? The Circuit Court must perform this ministerial or administrative duty upon the record of the evidence before it. If this Court can't find the Cleminshaw appraiser's value to be accurate because of the motive which influenced such assessment, there is not any reason to assume that 1/2 of such value would be twice as trustworthy or half as bad. In fact, mathematically, they are equally arbitrary, and must be given equal weight. None.

Would you believe it . . . the next year the dastard did it to me again. Back to court.

During the hearing I was asked to disclose all of my methods of finance. The county prosecutor said, "Yes, but you are forgetting something." I answered, "Could be. There is a lot of dimes changing hands out there. Refresh my memory and I will tell you." The prosecutor said, "How about the airplane and car mechanic business." I replied, "Oh, that does not exist anymore." Mr. Summerfield, the prosecutor, said, "Since when?" My answer, "Since you went into the car business at Oak Hill and hired my mechanic from me."

TAX

When the first copy of taxes was typed for the book, I discovered the secretary had the word dastard mispelled. She replied, "I thought I knew what you meant. I once worked for the county."

LETTER BY G. L. MASSEY

May 2nd, 1959

United States Senate
Washington, D.C.
Mr. Robert C. Byrd, Senator
Dear Senator Byrd:

Referring to the telephone call which I made in an effort to contact you. I do have a problem and I do believe you can be of help to me, and whether you can help me in this matter, I will still be forever grateful for you having allowed me the privilege of explaining the difficulty to you.

I have a small 1947 year model, J-3 Cub Piper Airplane, and have owned it since 1950. I bought the small airplane for my two boys to fly back and forth to V P I College at Blacksburg, Va. I also own and fly a 1957 Cessna 18C, which I bought new. But my problem lies around the Piper J-3 Cub. In 1955 I had this airplane recovered at the Dixie Aircraft Co. at Roanoke, Va. In 1956 I had the Engine, which was a 65 hp Continental, completely remanufactured making it into a 75hp engine. These two jobs cost me about 1166.00 dollars. The plane has about 10 hours on it since that work was done. In February 1957 I took this plane to the Huntington Airport, located at Chesapeake, Ohio, and left it for a 100 hr. inspection. The manager of the Airport in the following June, finally admitted that he had lost the papers to the airplane, but said he would apply for Duplicate papers at once. In October he said he finally had the plane ready for, and would have it inspected by a C.A.A. inspector. The Inspector was a Mr. Johnson, located at Charleston, W. Va. He refused to license the plane, and told me that he didn't believe those people would ever get it ready for a license. So I got a ferry permit from him and took it to Princeton, W. Va. and left it at the small airport there, where the operator said he could have it ready in about 30 days. At the end of thirty days I called the operator and he said it would be at least another thirty days, finally at the end of three months, I went to Princeton, to get the airplane and it was still parked in exactly the same place I had left it. And all this and the time it was at Huntington, it was parked outside in the weather and at my home I have a small airport and a very nice hangar to accomodate as many as three planes. That day I went to Princeton to get the plane, Mr.

Johnson was there again and so I had him give me another ferry permit and brought the Airplane back home with me.

I contacted a Mr. Harless, who is an A&E mechanic employed at Fayetteville Airport, owned and operated by Mr. Frank Thomas. He said he would be glad to license the plane. So, I told him to install all new glass, (plasticele) in all windows doors and windshield, to repair and correct all instruments, do a complete 100 hr. inspection, and to then call Mr. Johnson of the now F.A.A. to come and give it an inspection.

After the work was done and Mr. Harless said the plane was in as good mechanical condition as possible, he had Mr. Johnson come see the plane. He checked the plane and showed Mr. Harless some small spects of rust on some of the head bolts on the engine, and said that before he could approve the plane, the wings would have to be recovered and this rust would have to be removed.

So, I contracted the job covering the wings to Mr. Harless, plus cleaning the engine and painting it, for $320.00. I had already paid him $95.00 for the other work. Mr. Harles said he would do the job, but that he did not feel that the plane needed the recover. He also said that he didn't feel that there was another Cub plane in the country in as good shape as that plane.

The day that I called your office, Mr. Johnson had again inspected the plane. This time he cut fabric underneath the fuselage and found some small specks of rust on the cross brace of the frame structure. This rust according to Mr. Harless is of an exterior type and not a destructive type. Mr. Harless says that he will bet that Mr. Johnson can do the same thing and ground 90% of all fabric covered planes.

What I am trying to say is that I feel that Mr. Johnson is not being fair. I have talked with him and I find him to be a very indefinite type of person, who cannot talk an intelligent conversation with you at all. He wanders from one subject to another. I wish it possible for you to be able to interview him and see if you do not agree with me as to ability. You may use this letter in any way you like, because what I am saying to you will be verified by Mr. Harless and Mr. Frank Thomas, of the Fayetteville Airport.

Hoping you can help me get this matter cleared up, maybe just a telephone call to Mr. Johnson, would do the job, and hoping to see you again, soon, I am,

                        Respectfully yours,
                        G. LEE MASSEY

THOUGHTS OF 1978

It is not I as a person or a pilot with perhaps too much dedication to my profession, that there has been so much harassment from government groups. The pressure comes from those selfish gangs wishing to feather their own nest, those being the contractors, engineers, and politicians looking for big handouts from the government.

There are those among us who seek to destroy free enterprise through the pertinence of public need.

Our service at Fayette Airport has gone far beyond service existing at many of the public owned airports, which have cost the taxpayers millions of dollars.

In no way do I criticize the hard-working Flight Service Stations, Control Tower operators, those that man radar, searching the skies to help aid the pilots day and night. They are always there. Their duties are numerous, their *Quality* of service unequaled.

There is little I would change in the GADO office at Charleston if I could. They are bound by a set of rigid regulations. Often the Washington chief has not been a pilot. Sometimes he is a military man, perhaps believing nothing is good unless it is big and costly.

The Charleston FAA office has no resemblance to the dark days of aviation in West Virginia when we were in the Pittsburgh District office under Mr. Jones.

There was one year I was unable to establish one person receiving a pilot's federal license in West Virginia.

There is yet a small bit of favoritism to the municipally owned airports over the taxpaying airport. But things are getting better.

Some of the happenings in this book were nearly forty years ago, and may or may not fit today as they were during the 1950s to the middle 1960s.

I am too far down the road of life to fear what they will do to me or my flying school. My contribution to life is not this book, but my labor of love which was soloing nearly one thousand students.

My business of free enterprise is an endangered species. The young pilot should know what the old ones have endured. If we should return to the stagnation of flying we once witnessed, I shall again battle the *bureaucracy*.

ALPHA OR OMEGA

Through all of this, before, between and after each battle with the bureaucrats, I have reached an outstretched arm with a hand of friendship. It has not been grasped. I then continually made every effort to forgive and ask to be forgiven my mistakes. All I ask is to be treated as others. I request the government to stop trying to destroy free enterprise, and to treat the little man in business as he does those under their control.

The Dove of Peace is still in flight carrying an olive branch, looking for a place to land. If the Dove of Peace is not permitted to land and accept the olive branch, call this book Volume 1.

OMEGA

Brethren, I count not myself to have apprehended, but this one thing I do. Forget those things which are behind and reach forward toward those things which are before.

Philippians 3:13